Rice and Agricultural Policies in Japan

Nicole L. Freiner

Rice and Agricultural Policies in Japan

The Loss of a Traditional Lifestyle

Nicole L. Freiner
Bryant University
Smithfield, RI, USA

ISBN 978-3-030-08253-6   ISBN 978-3-319-91430-5   (eBook)
https://doi.org/10.1007/978-3-319-91430-5

Cover image: © sunabesyou
Cover design: Tom Howey

This Palgrave Macmillan imprint is published by the registered company Springer Nature Switzerland AG
The registered company address is: Gewerbestrasse 11, 6330 Cham, Switzerland

For the Morita Family, Shin Machi, Fukumitsu/Nanto, Toyama, Japan
Where "ただい"ま is always given a happy "おかえりなさい"
Thank you for the years of generosity, good cheer
and understanding warmth

Acknowledgements

This work developed out of a set of interviews completed back in 2013 with a Summer Stipend Award from Bryant University that provided me with the means to visit areas of Japan affected by the Fukushima Daiichi catastrophe and to meet with writers, mothers groups and non-governmental organizations doing work to bring attention to the many problems confronting people dealing with the aftermath of the crises. Through meetings and conversations that took place over a Summer Institute supported by the NEH at the East West Center in Honolulu, Hawaii on Buddhism, I was able to formulate a developing interest in the politics of food in general and rice in particular which brought me back to some of the concerns that were a part of my upbringing as someone who grew up on a dairy and crop farm in upstate, New York.

Several more Summer Research stipends provided by Bryant University allowed me to continue this work which had many contributions along the way from Japanese hosts who have always been beyond generous in giving of their time to answer my questions and help me to begin to understand a topic that has many complex and moving parts.

In particular, Steve Leeper, allowed me to stay at his farm where I was able to complete nearly 10 interviews over a weeks time and these early interviews were an important frame also informing my work on the Trans-Pacific Partnership. Mr. Mizukami and Mr. Saito in the village of Joge each spent a lot of time talking with me and explaining how government policies affected them. In Tokyo, the staff the the Citizens Union of Japan (Nishoren in Japanese) were very helpful as was its

Director Martin Frid who helped me to understand the politics of genetically modified foods and the perspectives of concerned citizens.

In Toyama prefecture, the local Nanto branch of JA (Japan Agriculture) met with me and also allowed me to tour the local country elevator which was a fantastic informative experience in getting at local level concerns, alllowing me to contextualize farmer's interactions with JA Zenchu and JA Zen-Noh. Sachiyo Abe arranged a number of meetings for me with local farmers and also assisted me in early translation and understanding some of the basic issues of local agriculture in the Fukushima/Nanto area.

I was also provided information by a number of people both at the Ministry of Agriculture, Fisheries and Forestry (MAFF) as well as JA-Zenchu, which helped me in answering important questions that came up in the early stages of my writing. Aurelia George-Mulgan, the person who knows the most about this topic because of the depth of her own research graciously answered my emails and helped me to develop my own perspective and confidence on the subject. Her work is in a category of its own making on the topic of agricultural politics and Japan, it has paved the way for future generations of scholars.

At Bryant University I would like to acknowledge the support and assistance of my Department Chair, Professor John Dietrich who has advocated on my behalf, allowing me much needed travel assistance for my many trips to conduct research in Japan.

I would also like to acknowledge my wonderful parents, Jim and Dianne Freiner who share my work with enthusiasm and joy. Though you may never read it in its entirety, you have had a hand in creating each page of this work.

Finally, I would like to acknowledge the loving support and generosity of my husband, Matthew Griffin who helped me to create my own special writing spot where mysterious gifts of chocolate and hot tea arrive at the most needed times.

CONTENTS

LIST OF FIGURES

LIST OF TABLES

Introduction: The Politics
of Rice in Japan

Introduction

The climate and topography of Japan are diverse, despite the relatively small scale of the country compared to its neighbors. With a total area of 377,962 hectares and an agricultural area of 4519 thousand hectares (roughly 3/4 the size of California), there are limits to the amount of available cropland.[1] Limits are also imposed by weather and growing conditions, especially in the mountainous regions, which account for 73% of Japan's topography. The climate ranges from severe cold weather in the Japanese Alps to tropical regions in the southernmost islands of Okinawa and the Ryukyus. To the north, the large open spaces of Hokkaido offer vast landscapes for planting and larger fields, their size (20 hectares or ha) is comparable to farms in Europe while in the rest of Japan the average size is less than 1 ha.[2] Hokkaido is Japan's biggest producer of rice and many other field crops, through breeding for cold weather tolerance, in the early 2000s Hokkaido accounted for 8% of Japan's total rice production.[3] As this book will explain, cultivation across Japan's landscapes is heavily influenced by the availability of water, a precious resource on this island for which a variety of traditional systems of sustainability have been devised. These include the communal irrigation systems that are used to drown rice fields during the early growing season in order to curb the growth of weeds without the use of expensive chemical fertilizers. These natural systems color the landscape, in one famous village, the spring water is channeled into homes and used for cleaning vegetables in pools, the water is cleaned by carp living in the pools and then channeled onward and distributed throughout the town through a network called *kabata*. This ingenious form of water control was once widespread but today, there are a small number of villages that use *kabata*. More villages, however, maintain the networks of water for rice growing. The communal management of resources in Japan is a distinct feature of agriculture; for rice, the management of water is the most important. Sharing of knowledge and resources is also evident in the agricultural cooperatives, institutions that are central to the agricultural experience in Japan, especially in the most rural areas.

Rice growing in Japan constitutes a complete life-world that is informed by state-directed forces as well as the pull of tradition felt in communities in rural areas and the centrality of rice to the Japanese diet. The way in which these forces interact makes for a fascinating story. It is my intention to tell a part of this story in this volume by bringing together different perspectives on the way in which rice growing can be understood, and explaining the interaction between the varied, sometimes competing forces acting on individual cultivators and growers. Today, the food industry, food cultivation, and the source of food are more salient topics than ever for consumers who, in the past several decades, have taken an interest in food issues and food politics. Here in the United States, the work of Michael Pollan and others opened up the conversation regarding food in popular culture, although it had already been an interest of researchers, especially those focusing on world hunger and food distribution since the 1970s. While perhaps popular interest in food was late to arrive, it is here, and already popular interest in food politics has changed the political landscape and is changing agriculture itself although it has yet to shift the global food industry to accommodate the interests of small growers or challenge the dominance of the largest producers.

Agriculture in many countries is an enterprise that is heavily regulated by the state, the connection between the way nation-states define themselves in nationalistic terms is still linked to agricultural production. In Japan, this is immediately apparent, the traditional Japanese diet or washoku is now on UNESCO's list of Intangible World Cultural Heritages[4] and the Japanese state has a policy directive of promoting this diet around the world in order to increase its rice exports. The agricultural experience holds sway over people's imaginations in powerful ways, not just in Japan but across the world, and it is on this landscape of imagination that one must consider the Japanese experience.

In the most basic way, this book is an examination of the issues of agricultural policy with regard to rice and seeks to understand the complex relationship between the Japanese government, its farmers, and intervening actors such as the cooperatives (or *nokyo*) and non-profit organizations that advocate on the behalf of consumers as well. This story is relevant and important to anyone interested in the role of government in maintaining family farming and the cultures associated with it, no matter where they are located. The response of national governments to international policy is also presented here because governments formulate national policy to meet up with the requirements of these agreements. Although it is not the focus of this book, the story of rice in Japan implicates all citizens and consumers of rice. Government policies may affect the price and availability of rice, which is the core of the traditional Japanese diet or *washoku*. Changing attitudes to food and shifting diets have already impacted the demand for rice and this is likely to continue, although the Japanese government is promoting the traditional diet with concerted attention. It is my hope that readers of this book come away with a broad understanding of the major issues and actors with regard to rice policy that includes the perspectives of farmers and citizen consumers.

RESEARCH SITES

The locations used for research sites in Japan include villages on the main island of Honshu as well as outer lying islands near Okinawa. In the south of Japan, I interviewed farmers in the villages of Yoge and Toyama as well as the island of Oita. These rural villages and those like them are the backbone of the farm lobby and the stronghold of the Liberal Democratic Party.

Joge is a remote village in Hiroshima prefecture, accessible via the Fukuen train line from Fukuyama from Osaka. The Fukuen line glides along the side of mountains with sheer drops on the other side. The town has a dwindling and aging population (mimicking the rest of the country), many of its young people have left for the cities and the remains of large rice farms sitting quietly is striking. Here, I met farmers willing to discuss the changes in rice farming at length to better understand how farmers and their lifestyle has been affected by recent policy changes by the Ministry of Agriculture, Fisheries and Forestry (MAFF) and commitments to multilateral agreements.

In the west side of Japan, in Toyama prefecture, I interviewed farmers located in the consolidated town of Nanto (formerly Fukumitsu) which sits in a valley near the Japanese Alps. This used to be called the backside of Japan, in an awkwardly arrogant manner by those living in Tokyo and other major cities. Recently, with the nearby Gasshou houses receiving recognition as a UNESCO World Heritage site, towns near the houses including Nanto are drawing artists and creative types. This part of Japan is traditionally conservative and like Joge is also an LDP stronghold with a farm lobby. Today, the entire western area of Japan is being promoted by the Japanese government as a place where young people in the cities can learn about Japanese culture, because it is one of the rare spots where that culture is still intact. I interviewed several farmers and village officials and members of Japan Agriculture or JA (the largest cooperative in Japan that is the focus of Chapter 4) here to learn about the impact of policy changes on the town and its economy.

Along with interviewing farmers in these research sites, I also spoke with representatives of non-governmental organizations in Japan representing consumer interests; government officials working for the Ministry for Agriculture, Fisheries and Forestry (MAFF), local representatives of the JA Zen-Noh or Japan Agriculture, the largest agricultural cooperative in Japan and representatives of the larger national organization JA Zenchu. These interviews were informative in understanding how policy changes are impacting these groups.

A qualitative methodology was used for this study, utilizing in-depth interviews in Japanese at the research sites above. The purposive sampling frame was used in order to focus on areas of Japan where farming was practiced widely and where farmers were members of cooperatives and were LDP strongholds. This sampling method was used because many scholars writing about the Japanese government posited a connection between the LDP and farmers, and this relationship is a major topic of this book. Although I set out to examine relationships and to suggest inferences based on this research, in no way is causality addressed. Moreover, the results of this research are not generalizable to countries outside of Japan. This study is only relatively generalizable to areas in Japan similar to those being examined.

ORGANIZATION OF THE BOOK

The book proceeds with Chapter 1, which discusses the role of rice in the Japanese diet and in the historical, cultural, and social traditions of Japan including early policies regulating rice and promoting self-sufficiency in order to ensure social stability. Here, the literature review is included as well.

Chapter 2 presents the context of policymaking in Japan with regard to the principle policymaking bodies affecting rice growers and agricultural policy affecting them. A general overview of the Japanese politics and the policymaking arena is presented before detailing those responsible for agriculture and rice policy. The principle body among these is the Ministry of Agriculture, Fisheries and Forestry (MAFF) which creates and implements agricultural policy. Also, the role of the LDP's agricultural policy tribe or norinzoku are discussed and the history of recent agricultural policies are presented.

Chapter 3 details the international arena of policymaking on agricultural issues, beginning with the General Agreement on Tariffs and Trade (GATT), the Uruguay Round on Agriculture and the World Trade Organization's Agreement on Agriculture (AOA). The way in which policymaking has progressed over time is presented along with details of these agreements and their potential impact on rice growers. This chapter also includes a summary of the Comprehensive Agreement for the Trans-Pacific Partnership and regional trade agreements, also explained are how measures included in the TPP would affect rice growers along with other important Japanese agricultural goods. Moreover, the status of the agreement and potential impact is also discussed.

Chapter 4 presents an overview of The Central Union for Japanese Agriculture or JA-Zenchu the national cooperative in Japan that advocates on the behalf of rice farmers and has also played a role in implementing MAFF policies including rice storage. Local level JA activities are also presented, which are coordinated by JA Zen-Noh, which oversees these local level branches of the Japan Agricultural cooperatives.

Chapter 5 discusses Japan's environmental and consumer movements and examines a number of the criticisms that these groups have put forward regarding legal changes necessary to adapt to international agreements such as the WTO AOA as well as the TPP. In particular, the issue of genetically modified organisms (GMOs) is detailed along with Japan's current legal regime regarding GMOs and the way in which they are treated by the aforementioned agreements. The response by these organizations and creation of additional citizens groups to address and manage problems regarding the impact of radiation on the food supply during the Fukushima Daiichi crisis is presented in this chapter as well.

Chapter 6 examines the issues of food security and food sovereignty, historical concerns of the Japanese government that have reasserted themselves in the past two decades. The impact of government intervention on the global rice market is discussed as well as the major features and background of the global trade in agricultural commodities in general and rice in particular.

In Conclusion, I will present my conclusions and after thoughts on the politics of agricultural policy in Japan with regard to rice, a special focus is given to the MAFF and the impact of its policies on rice growers as well as the global rice trade in general. The potential implications of international agreements and recent policy changes that adjust to such agreements are discussed in detail and suggestions for future researchers are also offered here. Moreover, the case of rice and policymaking on rice is analyzed within the context of larger debates about Japanese politics and criticisms regarding the democratic nature of the Japanese state and its policy interventions.

NOTES

1. UNFAO. 2018. "Japan Country Profile". Available at http://www.fao.org/countryprofiles/index/en/?iso3=JPN. Accessed June 21, 2018.
2. Ohara, Masashi. 2009. "Agriculture in Hokkaido." Available at https://ocw.hokudai.ac.jp/wp-content/uploads/2016/01/AgricultureInHokkaido-2009-Text-All.pdf. Accessed June 23, 2018.
3. Ibid.
4. UNESCO. 2013. "Washoku, traditional dietary cultures of the Japanese, notably for the celebration of New Year". Available at https://ich.unesco.org/en/RL/washoku-traditional-dietary-cultures-of-the-japanese-notably-for-the-celebration-of-new-year-00869. Accessed June 21, 2018.

CHAPTER 1

Japanese Rice: History and Cultural Performance

On December 4, 2013 the United Nations Educational, Scientific, and Cultural Organization (UNESCO) agreed to register traditional Japanese cuisine or *washoku* to the list of the World's Intangible Cultural Heritage list. With its entry to the list, it joined the traditional Mexican cuisines of Michoaca, the gastronomic meal of the French, and the Mediterranean diet, the only other cuisines to achieve this status. For the Japanese government, making the list was a status symbol, once again solidifying Japan's unique-ness. Registration on the list presumes that efforts will be made to preserve *washoku* and the Ministry of Agriculture, Forestry and Fisheries (MAFF) along with the announcement included statements regarding the marketing of Japanese cuisine to the world and a desire that Japanese themselves preserve and pass on the traditions associated with washoku for future generations.

The heralding of *washoku*, coincided with ongoing discussions in the Ministry regarding food sovereignty and the declining consumption of rice in the country. Rice is the central component of washoku, based on a menu of one soup and three dishes. Rice is the main dish and it is supplemented by the other three, as the MAFF Guidebook on washoku states "the purpose of the menu of washoku is to eat cooked rice with soup and side dishes" (p. 18).[1] Making the UNESCO list may afford Japan's MAFF more maneuverability as the country moves forward with international agreements like the Trans-Pacific Partnership (TPP); because rice is a bulwark of *washoku* it can be protected. Moreover, it gives the Japanese government the ability to market and promote rice

© The Author(s) 2019
N. L. Freiner, *Rice and Agricultural Policies in Japan*,
https://doi.org/10.1007/978-3-319-91430-5_1

as part of washoku to its own population. These policies are the focus of Chapter 2, where they are presented in detail.

RICE IN THE JAPANESE IMAGINATION

Rice is intimately connected to notions of what it means to be Japanese of Japanese-ness in Japan, both in the national consciousness and in national policy (a notion that emerged from the Meiji era specifically called *Nihonjinron*). These two are interrelated, as historians, sociologists and researchers note, nationalism is driven by policy and it is related to frameworks of identity that are established by governments rather than existing in a separate hypothetical space created by citizen's imaginations. In Japan, while rice is a recent historical tradition, there are references to rice in Japan's earliest historical writings, including the Kojiki and Nihon Shoki. Ohnuki-Tierney describes a gradual process whereby Japan's agrarian cosmology and rituals associated with agriculture became the bulwark of the imperial system in her book *Rice as Self: Japanese Identities Through Time*[2] the only book written on the cultural importance of rice in Japanese culture. The book sets out to examine the importance of rice in understanding notions of Japanese identity and cultural performativity, which the work accomplishes with a focus on historical understandings as well as ritualized cultural practices, the role of the Emperor and the centrality of rice the Japanese diet, at least as it exists in the minds of Japanese people. She also describes the role which agrarian harvests played in the leader's ability to hold onto power, which is similar to Chinese belief in the relationship between their divine ruler and an abundant harvest.

In Japan, the annual harvest ritual legitimated political leadership which served as the officiant in rituals for the rice soul (*inadama no shu-saisha*). Early rituals (called *matsurigoto*) that were the basis of the political system (*ritsuryou sei*) were all related to the rice harvest and so power was associated with rice. Despite the cultural association that exists currently between rice growing and ideas regarding Japanese identity and national policy; the culture of growing rice in small paddies is a fairly recent phenomenon in Japanese history.[3] However, this linkage continues to hold sway over Japanese farmers and rice growers, as well as the traditional ruling elite in Japan (the more conservative wing of the Liberal Democratic Party of LDP). How did rice come to play such an important role in Japanese culture? And, how will this association hold up as Japanese diets are changing and fewer and fewer traditional rice growers live the rice growing lifestyle? This book sets out to begin to frame answers to some of these large questions. While it may not provide a definitive and complete answer, I hope to at least tell part of the story, from the perspective of rice farmers, their families and the rural areas where more and more abandoned rice paddies are growing weeds. As a farm child myself, I understand the power of cultural associations with the family farm and the impact that changing diets, new international trade relationships and concerns about food sovereignty have upon the existence of family farms which have become almost extinct in the United States.

There is a fascination with the family farm in both American and Japanese culture. The family farm is the cultural repository of wholesomeness, of an almost idyllic existence which probably beckons to our framing of the natural world itself and the myth of an Eden like paradise where humans once lived in harmony with each other and their environment. These seemingly lofty ideas have real political impact, as governments have created policies to preserve family farming and the cultures associated with this lifestyle around the world. While this book does not compare the United States and Japan, the researcher is American and therefore some comparison is implied because my writing is framed by my own experiences. At the very least, I would like to illuminate a lifestyle and document what is happening in Japan today, as rice growers are giving up the farms or simply dying out and taking the narrative of their lives with them, and their experience as rice growing to the grave.

SITUATING THE ARGUMENT OF THE TEXT

One would not suppose, given its population and lack of arable land, that Japan would be a country devoted to agriculture. Yet, Japan is a country with a high degree of political and social interest in protecting agriculture, some of the reasons for which have been discussed above. Japan's policies do not treat all crops equally with rice being the crop where devotion is singularly manifested. The policy framework regarding rice is relatively new, with patterns inherited from the pre-WWII period of the 1920s and 30s after a series of bad harvests resulting from drought amidst a brutal tax policy brought famine and political unrest. The current framework, though significantly altered after land laws were reformed post-WWII owes its geneological heritage to this timeframe. This book addresses the major actors in this policy framework, which includes public and private actors, laws, legal norms and policies as well as informal social currencies. The interests and activities of these actors connect and overlap, sometimes diverging, sometimes harmonizing, and despite recent attempts to overhaul them, their strength resonates to local level farmers and their families. The research which examines rice policy from the perspective of farmers in Japan is scant, however, there are a number of well-researched books and articles on agricultural policy with rice as their focus. Foremost among this research are the two volumes written by Aurelia George-Mulgan, whose highly informative and pathbreaking work covers many aspects of agricultural policy. In Mulgan's first volume,[4] the focus is on the Ministry of Agriculture, Fisheries and Forestry and the interventionist manner in which it has maintained its role in agricultural policy even advocating policies that are to the detriment of the progression of farming in Japan in order to keep power and continue to draw large sums from the Japanese budget. Mulgan's work also covers the internal dynamics of the MAFF and the close relationship of its bureaucrats work LDP members of the agricultural policy tribe or *norinzoku*. No other author approximates the highly detailed nature of Mulgan's research, which if taken as a whole is a lifetime of work shedding light on the agricultural policy arena in Japan, especially the MAFF. Her second volume[5] on agricultural policy presents the context of agricultural policymaking in Japan at the national level, essentially building on the first volume and moving the focus of analysis to a wider discussion that provides the reader with a deep understanding of the national level policy actors and political changes that have informed policymaking. Taken together these two volumes provide a highly detailed analysis of the agricultural policy regime and its most important actors, the MAFF.

The literature on rice farming in Japan in the field of political science is sparse, there is no single study covering the politics of rice growing from the perspective of farmers and the studies of the politics of agriculture are rare. Yoshihisa Godo has written several articles that examine rice farming and the policies which have maintained it with detailed descriptions of the policies and their backgrounds. Godo's detailed background work on policy examined alongside Mulgan's work provides an excellent frame within which to understand the basic components of agricultural policy, the role of the MAFF and the overall landscape of rice policy.

Land reform and the importance of land reform for landownership following WWII is the topic of Dore's, *Land Reform in Japan*,[6] another work which is pathbreaking. Unfortunately, it is the single publication covering these issues in any depth; it has not been followed up by complementary research in the modern or current time frame. The work, however, provides important background and context for agricultural policies and landownership, without which, there would likely be few farmers for those policies to represent. Waswo and Yoshiaki, in their text *Farmers and Village Life in Twentieth Century*[7] update some of the information in Dore's work. Their edited volume presents episodic portraits of the lives of people in rural Japan and some information regarding rice farming and the farmer's perspective is hinted at although it is not the focus of the research.

Researchers on Japanese politics have noted the centrality of the Japanese state, where power is located mainly in national level institutions. In this policymaking context, the Cabinet and their leaders are important initiators of policy, in previous studies, authors have advocated two basic arguments about the organization of the Japanese state. One camp of writers, has argued that Japan's bureaucracy is an "iron triangle" with a high degree and overlap between members of the LDP, Japan's most powerful political party with bureaucrats and high level business interests. The iron triangle or elite power argument has sustained a devoted following among scholars of Japanese politics but it has been criticized for its focus on vested interests and coordination across groups that most times have very different interests. An alternative argument is presented by scholars who note the role of factionalism in the LDP and the intra-bureau conflict which prevents these groups from behaving in the manner supposed by elite power scholars.

The arguments of these two schools of thought have been supplemented by more recent scholarship that goes beyond the national level studies to illuminate the inner workings of the bureaucracies and also include grassroots level actors. These include several studies that present arguments regarding the interventionist nature of the Japanese state. The main thrust of this argument, used by Sheldon Garon[8] and George-Mulgan, discussed previously, is that the common feature of the Japanese bureaucracy is its intervention in the Japanese economy and society through an array of institutions and legal frameworks that seek to maximize the power of the individual ministries to intervene through these institutions to preserve and consolidate their power. At times, these interventions direct social policies and absorb civil society actors as well.

An additional set of literature on Japanese politics seeks to illuminate the activities of civil society actors such as environmental groups, women's movements and others that have often been diminished or underplayed by scholars. Robin Leblanc's *Bicycle Citizens*[9] for example is one such study which argues that the political activities of women occur in ways that are not captured by theories of democratic politics whose assumptions about the conflictual and vocal nature of citizen's groups leads scholars to miss the activities of some local level, grassroots actors that do politics in methods that are judged as passive or cooperative by democratic theories. Increasingly, studies of civil society actors and their responses have become more nuanced, this work is exemplified in the book on the aftermath of the triple disasters edited by Tom Gill et al., *Japan Copes with Calamity*[10] a detailed ethnography that includes activity by youths, women and local level research. Other examples of such works also include Aldrich's study on the siting of nuclear facilities and citizen protest, the presentation of the movements to address Minamata disease by Timothy George,[11] and Michael Lewis' book on citizen protest at the local level in Toyama prefecture.[12] The insights provided by these authors add depth to the existing literature on Japanese politics which has a tendency to focus on the national level of government. The story does not end there, however, local level actors manipulate policies and this growing body of literature has begun to address these questions although more work needs to be done.

On the relationship of Japan to the world with regard to global trade, Kym Anderson's[13] recent work illuminates the way in which farm policies affect developing countries and provide one with an understanding of the issues involved with agriculture in the World Trade Organization's Agreement on Agriculture and other agreements. Christina Davis's work on food, trade liberalization, and international institutions is extremely insightful in laying out the major issues at play and the interaction of international and domestic level institutions. The literature on global trade is very rich with regard to Asia and Japan, there are a number of well-researched volumes on these very complex relationships that also reveal Japan's role in international trade, its policies and assist in understanding the trade strategies it uses. Hayami and Godo add to this narrative with their detailed work on the role of rice policy in Japan's domestic politics as well as the interaction between these two forces.

There are two sets of literatures on Japan's domestic policies and relationships that are especially sparse, One is work on issues around genetically modified foods and Genetically Modified Organism (GMO) policy. Another is in-depth longitudinal studies of environmental movements and related citizen protest. There is no work at present that analyzes the GMO movement in Japan or the policies and institutions that exist with regard to GMOs and as yet there is also no long-term analysis of the various citizens and consumer groups motivated by the environment, health, and consumer issues. Although some of the studies mentioned previously look at the issue of citizen protest, they are all discrete, focusing on particular movements that respond to a specific problem or situation. This is very unfortunate as scholars don't have a comprehensive overall view of how these organizations fit into the larger story of Japanese politics. On issues of global environmental policy including issues related to food sovereignty and security, Lester Brown[14] stands out as author who is prolific, his volumes are well-researched and highly detailed, he provides the reader with an understanding of oftentimes very complex relationships with a focus on how these relationships matter.

On issues of food security and sovereignty in Japan and Asia, several volumes stand out, including the volume by Josling et al.[15] on global food regulation and trade which provides an excellent overview of what is at stake with regard to the food regime and the lack of a rules-based system that provides oversight. In two excellent volumes, Brown[16] provides excellent details on the numbers and statistics with regard to global food trade and contextualizes that information within global climate and environmental changes. The picture that Brown paints is sometimes dire, and his predictions are stark, especially for Asian countries dependent on others for staple foods but the entire world and its consumers will be affected by the interplay of the forces that he describes.

This book's title suggests that changes in agricultural policies will, or are, having a negative impact on a set of cultural values in Japan associated with agriculture. The definition of traditional therefore implies a set of relationships that at their core are also connected with nationalism. The communal practices associated with rice growing are relatively recent in Japan's history, as the next chapter details. The main crop, rice and manner of growing it was changed in a revolutionary manner, with the introduction of dwarf rice and chemical fertilizers in the 1960s. Powerful incentives driven by political institutions have shaped the actors associated with rice agriculture and these incentives occur in the context of global agricultural supply and demand and trade. The global trade order established after WWII, with the Bretton Woods institutions and the General Agreement on Tariffs and Trade started a process of constricting state sovereignty that has intensified with the World Trade Organization. This book focuses on examining these relationships with a view toward farmer's, citizens and community concerns. These concerns take place within a social construction of reality and practices that have become ritualized and tied to the way in which the Japanese nation-state imagines itself. The agricultural policy framework in Japan illustrates the cooperative activity that takes place between private social actors and formal policymaking structures, noted by authors such as Verba, Nye and Kim[17] and Dittmer, Fukui and Lee.[18] Therefore, the way that agricultural policy is implemented and formulated, is influenced by farmer's organizations like Japan Agriculture which has channeled government funding through its programs and at times acted as a quasi-governmental body because of how deeply it is involved in all aspects of rice growing, marketing, distribution and storage. Rice growing is a part of the Japanese nation's self-conceptualization, as such, it plays a performative cultural role in attending to these understandings. This book argues that the maintenance of rice growing, in its current form in Japan is about more than bureaucratic self-preservation and the Ministry of Agriculture Fisheries and Forestry's entrenched interests, it is about the Japanese nation-state itself. Overlapping interests which are now disentangling themselves, the conservative nature of the Liberal Democratic Party and its current leader, Prime Minister Shinzo Abe have been fundamental forces in shaping these interests and hastening the demise of the agricultural policy framework which has been in existence since the post-WWII era. This book will demonstrate the importance attached to this cultural role, by examining the policies of Japan's agricultural policy structure and the current policies which these institutions are pursuing.

History of Rice Growing and Government Policy in Japan

During the modern era, the beginning of Japanese government manipulation of the supply, trade and price of rice in earnest goes back to 1918, when the Rice Riots of 1918 broke out. During the interwar period and prior to this time, rice was a luxury good for many across Asia, produced by those indentured to large landholders to pay the tax demanded by feudal lords (or bakufu幕府) but relying on other grains for their everyday meals (Table 1.1). The rice taxation system or *kokudaka* was based on the putative yield of rice in each lord's territory. Governments, temples, and shrines provided peasants with un-hulled rice seeds for spring

Table 1.1 Periods in Japanese history

Period	Dynasty	Year
Prehistory	Jomon	10000BCE–300BCE
	Yayoi	300BCE–300CE
Ancient	Yamato	300–700
	Asuka	592–710
	Nara	710–784
	Heian	794–1192
Medieval	Kamakura	1192–1133
	Muromachi	1336–1568
Early Modern	Azuchi-Momoyama	1568–1603
	Edo	1603–1867
Modern	Meiji	1868–1912
	Taisho	1912–1926
	Showa	1926–1989
	Heisei	1989–present

planting. Then in fall, after harvest time, the seed loans were repaid with a new crop of rice. This practice, known as *suiko* was the center of the barter system in which the use of rice was preferred over cash payment. Rice was the preferred medium of exchange, considered pure whereas metallic currency was considered impure.

The belief that metal currency is "dirty money" persists in Japan even today, although it is disappearing. Still, in most small towns, cashiers use a small plastic dish to hand metal currency to a customer for change rather than touching hands and cashiers in larger cities hand metal currency to customers using the receipt as a barrier between the change and the customers hand. For special occasions, rice is still the medium of exchange and is the most important offering to both the home Buddhist alter and Shinto shrine.

Prior to the modern period, in Japan's Medieval and Early Modern periods, most families tended their own rice paddies and the tradition was for each family to keep their own rice seeds for next year's planting, a practice that was endorsed and protected by the government (Tierney 2004).[19] During the Medieval era, rice was grown in small fields by peasants working fields to pay tax and provide for the family. Fields were measured in chou (one chou is about 2.5 acres) and a good field would produce about 10 koku (5 bushels of dry measure), with average annual consumption of one person being about one koku. There was a great freedom of movement during the Middle Ages along with social mobility, until Toyotomi Hidoyoshi completed his great cadastral survey fixing people in their status and to their soil. His intention was to fix farmer soldier and disarm, an announcement made after the survey was completed and all soldiers were told to disarm unless they were of the military class.[20] The bakufu focused on a policy to tax the peasants to the point of exhaustion, they lived a wretched life without the hope of change. Even during the Tokugawa regime, government oversight of crops was strenuous, with supplies closely monitored because of potential shortages.

In 1733 a number of riots broke out because a pest infestation in western Japan diminished the crop, leading to stricter control of the Rice Exchange which was the heart of the rice market. The price of rice rose so high that authorities were unable to devise a remedy, there were serious riots against speculators trying to corner supplies. Government authorities stalled to create a solution and intervened by fixing an official price in 1735, the first time this policy was used. In a country prone to typhoons, varying feudal kingdoms that the government had little control, the regimes search for stability (especially stable currency) relied upon the rice tax. As Sansom notes,[21] during the reign of the Bakufu (Tokugawa shogunate) "(S)ince the peasants provided the staple food of the whole country it was essential for the government to keep control of agriculture" (p. 104).

Control meant intervention in village life along with steering the Rice Exchange. The most important intervention was to use the same method as during the population survey, which was measuring and assessing the product of farms and taxing them. "The procedure involved a close examination of the land its yield by inspecting officers who were on the look-out for lazing farming" (p. 104). One can scarcely imagine a more coercive measure of production farmers.

During the Modern period, the government took over where the shogun left-off and used rice as part of a rice control policy in order to provide inexpensive rice to workers, to keep living costs down and justify low wages. The government took over distribution in earnest in 1939 subsidized rice through payments for rice production and for farmland improvements. Today once again as discussions of rice self-sufficiency are being made by the Ministry for Agriculture, Fisheries and Forestry, rice growing and policies associated with it are part of efforts related to political power.

The involvement of the Japanese government in rice and rice policy is unparalleled in its degree and unmatched in the world. The reasons for this are related to the manner in which rice has been consistently linked to the political narrative of nationalism that dominated the Meiji era efforts to promote a coherent notion of national identity in Japan. After the government's consolidation during the Meiji Restoration, a time that Japan also experimented with philosophies of democratization and the promotion of a unified central government, the government took over where the shogun left-off. Rice then was never part of Japan's market economy, rather the supply, distribution, and sale of Japan's predominant staple or *shushoku* was managed by national level bureaucrats.

During the Meiji era period of state consolidation, the Ministry of Agriculture and Commerce (MAC), the ministry responsible for formulating policy on areas such as rice production, was divided and policies were problematic and contradictory. This rift in the MAC because of its dual mission meant that the people responsible for agriculture (*Nomukyoku*) and those responsible for industry (*Shomukyoku*) had different goals. The Nomukyoku was responsible for land improvement programs and production, it was led by a Minister who wished to preserve small-scale cultivation and their supporting households "as the backbone of the country, a morally and physically superior to urban producers."[22] Rice cultivation was then, linked to the new nationalist narrative which was emerging in Meiji era Japan.

Mass-market consumption and the initiation of rice as the food staple began in the modernizing period (roughly 1890–1920s) when the first phase of industrialization began in Japan. These rice protests were perhaps the most serious and violent incidents of civil disturbance in the modern era. As Lewis argues, the significance of the riots was multifaceted, some enduring and others providing temporary relief.[23] One major outcome was the government's adoption of the Imperial Self-Sufficiency Policy. This policy was designed to address rice shortages and to ensure national food security that had been demanded by various groups petitioning their representatives (Lewis 1990). Additionally, the majority party "began to put the interests of cultivating tenants and urban consumers before that of landlords" (Lewis 1990, 245).[24] As a result, the MAC began a reclamation plan aimed to increase the land available for planting, encouraged uniform paddy sizes and financed the building of local rice warehouses (Lewis 1990) (Table 1.2).

Table 1.2 Policy response to the rice riots

	Year
New laws	
Rice Self-Sufficiency Act	1918
Rice Control Act	1921
Poor Relief Law	1929
Health Insurance Act	1927
Policy changes	
Independent Ministry of Agriculture created[a]	1918
Volunteer Welfare Commissioner System	1918

[a]The Ministry of Agriculture and Commerce existed prior to this change

Part of Japan's self-sufficiency policy involved the annexation of Korea in 1910 and Taiwan (known then as Formosa) in order to have additional land for rice growing. A 20-year program was started to eliminate rice supply shortages in Japan by supplementing Japanese grown rice with rice from Korea and Taiwan. The program "provided for the irrigation and drainage of large tracts of land... the introduction of commercial fertilizer and of modern scientific farming methods, and the substitution of Japanese varieties of rice for the native varieties (in both Korea and Taiwan) formerly grown there (IPR 1933, 1)."[25] While this program and the larger policy directing it were aimed only at rice supply and ensuring the ability to meet domestic demand for rice, over time the focus of the government's efforts shifted from supply to price. Very quickly, the amount of rice imported from both Korea and Taiwan grew rapidly while domestic production remained relatively stable; along with these changes was the growth in a group of domestic growers who now had an interest in opposing price controls. The Rice Control Act of 1921 was legislation intended to address supply issues and allow the government to purchase rice to support the domestic market in case of a severe price drop. The law permitted variation in tariff rates and government purchases and sales of rice to maintain a stable food supply. Another effect of the rice riots was related to long-term planning and stepping up social welfare programs. These programs were intended to provide relief in times of crisis (such as the rice shortages) especially at the local level. In prefectural governments across Japan, welfare agencies were created and the term "social welfare" began to become a part of the state's language.

While production was steadily rising (especially in Korea and Taiwan) in the United States the Great Depression hit in 1929 and had a global effect. The price of rice began to drop severely prompting landlords and farmers to pressure the government for price support. The Rice Control Act was revised in 1931, 1932, and 1933 to such an extent that in essence each time it was a new piece of legislation.[26] These revisions introduced the embargo of all foreign rice and created a standard for price control that defined a range for the price of rice, outside of which, the government would make adjustments (IPR 1933).[27]

Even this early in the Japanese government's relationship with rice farmers and producers the measures were politically controversial and highly costly (in the 1930s the cost was approximately 700 million yen per year). However, a political lobby also resulted between landowners and the tenant farmers that grew rice as well as the farmers who owned their own land and grow their own rice. When land tenancy relationships changed (a topic to be discussed further in this chapter) this politically powerful collusion would alter, but the rice farm lobby would lose none of its power. In 1925, the MAC was split into two separate bodies: the Ministry of Commerce and the Ministry of Agriculture.

Following this split, the Ministry of Agriculture's independence allowed it to "take over control of agricultural price and trade policy and to run it in tandem with policies for expansion of domestic output" (Francks 136).[28] Despite the separation of the two ministries there was tension between them and while they were fused again during the war years, after the war they were once against separated and renamed. The Ministry of Agriculture became the Ministry for Agriculture, Fisheries and Forestry and the Ministry of Commerce because the Ministry of International Trade and Industry.

In the postwar era, the bureaucracy was unwilling to alter its treatment of rice as a national treasure, maintaining high levels of bureaucratic support and a small-scale labor-intensive system of rice cultivation which failed to become industrialized or commercialized as other major parts of the economy were. The first ten years after the war, Japanese farmers received prices slightly above the international average, because prices were kept close to international levels but per capita supplies were less than 3/4 of prewar levels. Demand elevated the price and even created a black market. From the end of the war to 1948 the average income of farmers, far exceeded the national average income as a result of a rising demand for rice (with a large population of repatriated people from occupied territories and high international grain prices).[29]

The Food Control Law of 1942 set prices and raised them periodically to encourage deliveries as well as providing production incentives and subsidies. After 1948, the farm income was eroded for a number of reasons. Firstly, nonfarm incomes grew more rapidly and reconstruction increased the demand for labor in construction and factories. Secondly, the Korean War helped to eliminate the postwar black markets (including the one for rice) that lowered prices for rice. Overall, exports and imports of all goods become more stable as a result. The Land reform laws imposed a limit on the size of large farms and landholdings which, along with increases in nonfarm income, meant that small farms increasingly earned less. By 1960, farm incomes were well below the incomes of other households even though off-farm earnings had increased. This led to strong demands for protection on the grounds of equity and social justice. The combination of overrepresentation of rural voters and strong support for conservatives in the countryside has ensured a high responsiveness on the part of the LDP to these rural clients. In addition to influencing government through local representation in the diet, farmers have also in the postwar era had a network of organizations to assist them.

The Agricultural Cooperative Union Law of 1947 established a breed of cooperatives called *Nogyo Kyodo Kumiai* or nokyo (野狂). Through federations at national and prefectural levels these individual co-ops create a close-knit nationwide organization to which virtually all farmers belong. Since the 1970s, farm membership in *nokyo* has exceeded the total number of farm households. It is the largest voluntary grouping in Japan. Activities of these cooperatives extend into every aspect of farm production, including welfare, social, and cultural needs of the agricultural population. The bulk of farm produce is marketed through the cooperatives, especially rice and other cereals, while farmers purchase most of their farm requisites and a smaller proportion of their household needs through the same channels. These cooperatives are in no way new. During the prewar years the government attempted to foster cooperatives for tenant farmers but commercial and industrial interests blocked these efforts. After the war, when land reform became a reality and former landowners lost their control over large parcels of land, tenant farmers were able to become independent. With the independence of former

tenant farmers, the government was able to strengthen these organizations, albeit under new nomenclature, which during the prewar years it
had "loaned upwards of one hundred million yen at nominal or no interest" to alleviate the poverty of tenant farmers (Fisher 1937, 2).[30]

During the prewar years, these organizations had also begun to
achieve some political successes as farmer federations "adopted ringing
platforms, calling for such objectives as cheaper hydro-electric power,
protection against eviction, nationalization of fertilizers, revision of the
Civil Code to give status to the unions and freedom of assembly and agi-
tation..." (Fisher 1937, 2).[31] Before the war, the cooperatives supported
by farmers were also having success at the national level, though it was
tentative. Fisher states that farmers helped to double the diet strength
of the Social and Masses Party in 1936 and 1937. While it is impossible to speculate, certainly the farmers were on their way to becoming
a political force under the Tokugawa regime. Between 1955 and 1970
as incomes rose and the use of intensive high-skilled labor meant that
the country was not dependent on low wage labor, average income more
than tripled and real wage rates more than doubled. The share of rice in
household spending dropped sharply from 13% in 1955 to 4% in 1970.[32]
Much of the explanation for the rising levels of protection during the
postwar period have to do with a downward shift in the supply curve for
that policy, reduced opposition from nonagricultural interests lowered
the political cost to the government of supplying that policy. Taxpayers
and consumers as well as trade interests and less incentive to oppose the
demands for increased agricultural protection and the share of agriculture in the national income decreased, making it less burdensome for the
rest of the population to shoulder the cost of that burden. Moreover,
many of Japan's rural population feel a kinship with their farming relatives and the rural population, which has a disproportionate share of
votes is important to the LDP (Hayami 1982).[33]

The policies through the 1950s were based on the idea that farm
incomes and nonfarm incomes should be relatively equal. But as farmers incomes grew above those of nonfarmer, the government introduced a new formula. The new formula took cash and non cash costs
and provided for paying farmers work at rates comparable with those
of industrial workers in provincial cities. The Agricultural Basic Law of
1961 enshrined this policy as a fundamental principle. During the 1960s
and through the 1970s, rice consumption declined steadily and the

government adopted plans to reduce the acreage (*gentan*) devoted to rice growing. The Acreage Control Program or *gentan* began in 1970, it was intended to promote the diversification of crop growing, away from rice and toward other staples. However as Yoshiaki (2003)[34] illustrates in rural areas this program had the unintended consequence of making many farmers focus more intensively on rice growing while others left farming completely. This doubling down can be explained by the fact that growing rice was still more profitable than other crops (because of other government programs which controlled the price of rice, allowing farmers to profit). Moreover, farmers that were growing rice could hardly manage the additional time needed to grow other crops successfully along with rice, which took less time than other grains to grow and maintain (Yoshiaki 2003).[35]

During Japan's spectacular postwar economic development, the farm sector retained a demographic social and political importance far greater than its economic importance to the nation. The ability and willingness of nonfarm groups to tolerate the budgetary costs of protection have wavered at times, but in general Japanese people have favored protection. Japanese people view the costs of agricultural protection as an insurance premium against the possibility of food shortages arising from any future breakdown in Japan's access to imported supplies, or they view the farm sector as a residue of traditional Japanese culture and want to protect it (Mulgan and Saxon 1982).[36]

A key feature of farming in Japan noted by several authors (Francks 2000; Jussaume 2003; George 2001)[37] is part-time farming or pluriactive farm households. In her work, Francks notes that part-time farming is one of three similar features across Asia during the industrial development phase. Understanding the role of part-time farming is useful because it helps us to examine the role of community and the link to farming which these households sought to maintain. Rather than simply giving up farming to move to industrial areas, farmers adapted to industrial change by becoming part-time farmers. As Jussaume (2003)[38] notes, most rural households in the 1940s earned their income from farming but by the end of the 1960s, most rural households gained more than half of their income other sources. Part-time farmers have at times been criticized by the popular press for being beneficiaries of extremely costly government policies that have subsidized the part-time incomes of rice growers in particular allowing them to maintain the farming lifestyle.

LANDLORD TENANT RELATIONSHIPS AND LANDOWNERSHIP

Prior to the industrial area, during feudal times, farming was the basis of the Japanese lifestyle, along with fishing. Moreover, most Japanese did not own land and were subject to the demands of payment (in the form of wheat, rice, and barley) to the local *daimyo* (feudal lord). By the late 1900s, after the sweeping changes of the Meiji Restoration, many of those residing in rural villages owned their own land or were tenant—owners dividing their time between working their own land and working for a larger landowner. In many rural areas across Japan, landlord–tenant relationships remained although they had begun to change. These relationships have been illuminated by previous researchers such as Yoshiaki (2003), Waswo (1989) and in literary accounts such as Nagatsuka Takashi's famous novel *Tsuchi* (the Soil).[39] During the WWI, economic growth in Japan was rapid and the state was able to take on a number of projects to make life easier for its tenant farmers.

A large-scale project of land adjustment (*koucho seiri*) in local communities transferred large parcels into smaller uniformly sized parcels, allowing landowners to exchange holdings with one another so that their land was less scattered than in past. In Japan, the traditional model of landownership is based on communities. Individual farmers own parcels that are part of larger fields, but their individual parcels were likely to be scattered about in the larger field. This pattern necessitated the communal management of water resources, encouraged communal planning for growing, and also pressured individual farmers to take good care of their plots. The growth in agricultural productivity assisted by the use of fertilizers (such as soybean cakes) increased the amount of land devoted to rice growing significantly. Yoshiaki (2003)[40] illustrates that "between 1915 and 1924 the average annual volume of traded rose to 309.25 million koku, 5.38 times the average annual output of 57.59 koku" (p. 14). The lives of farmer began to change and this once weak economic group, living consistently on the edge of destitution began to become community producers. Now experiencing a taste of economic freedom, the stage was set for an increase in landlord–tenant disputes as farmers sought a more equitable distribution of economic benefits and to see the wealth of their output instead of handing it over to a landlord.

As discussed earlier, the Rice Riots of 1918 altered the agricultural community in Japan in dramatic ways that laid the groundwork for future policies committing the Japanese state to self-sufficiency in rice and a number of policies that protect the livelihoods of farmers. The landlord–tenant disputes during the 1920s and 30s resulted in a politicization of rural farmers who increasingly joined associations to pressure landlords for rent reductions and many also began to represent their own interests in village politics and local affairs. However, the Showa Depression of 1930–1931 would put a damper on this forward momentum. Many farmers were indebted to the landlords and asked for modest rent reductions during the Depression throwing the tenant farmers movement into a weaker position. By the late 1930s, rural areas had begun to recover from the depression and because of the important role they played as producers during the 15 Years War, many farmers were able to pay off their debts (Yoshiaki 2003).[41] A number of laws were passed protecting the rights of farmers, including the 1939 Farm Rent Control Ordinance, the 1941 Emergency Measures for the Management of Farmland and Special Control Ordinance on Farmland Prices, the 1939 Farmland Adjustment Law and the 1942 Staple Food Control Law. By far, the most important of these measures was the Staple Food Control Law, detailed earlier which provided the Japanese government with the ability to control the price and distribution of rice.

The Rice Riot, Showa Depression and the policies initiated to address them would provide the framework for the national policy apparatus that would begin to take shape after WWII. Foremost among these changes are the land reforms that dissolved large landholders' estates making it possible for former peasants and poor village tenant farmers to own their land. These groups would become the backbone of Japan's agricultural production during the highpoint of Japan's economic growth through the 1980s as well as forming the basis for the LDP's foothold in rural villages that until recently was the LDP's most important political constituency. The continuing policies regarding agriculture and rice growing are presented in detail along with the institutions responsible for implementing them in Chapter 2, which follows.

Notes

1. Kumakura, Isao (ed). *Washoku: Traditional Dietary Cultures of the Japanese.* Tokyo: Ministry of Agriculture, Fisheries and Forestry (MAFF). Available at http://www.maff.go.jp/e/japan_food/washoku/. Accessed December 3, 2017.
2. Ohnuki-Tierney, Emiko. 1993. "Chapter Six: Rise as Welf, Rice Paddies as Our Land". In *Rice as Self: Japanese Identities Through Time*, 81–98. Princeton, NJ: Princeton University Press.
3. See Ohnuki-Tierney (1993), Dore (1985), and Francks (2003).
4. George-Mulgan, Aurelia. 2005. *Japan's Interventionist State: The Role of the MAFF.* London: RoutledgeCurzon.
5. George-Mulgan, Aurelia. 2006. *Japan's Agricultural Policy Regime.* New York, NY: Routledge.
6. Dore, R. P. 1985. *Land Reform in Japan.* New York: Schocken Books.
7. Waswo, Ann, and Nishida Yoshiaki (eds). 2003. *Farmers and Village Life in Twentieth-Century Japan.* London: RoutledgeCurzon.
8. Garon, Sheldon. 1997. *Molding Japanese Minds: The State in Everyday Life.* Princeton: Princeton University Press.
9. Leblanc, Robin M. 1999. *Bicycle Citizens: The Political World of the Japanese Housewife.* Berkeley: University of California Press.
10. Gill, Tom, Brigitte Steger, and David H. Slater, eds. 2013. *Japan Copes with Calamity: Ethnographies of the Earthquake, Tsunami and Nuclear Disasters of March 2011.* Bern, Switzerland: Peter Lang Press.
11. George, Timothy S. 2001. *Minamata: Pollution and the Struggle for Democracy in Postwar Japan.* Cambridge: Harvard East Asian Monographs.
12. Lewis, Michael. 1990. *Rioters and Citizens: Mass Protest in Imperial Japan.* Berkeley: University of California Press.
13. Kym Anderson. *Finishing Global Farm Trade Reform: Implications for Developing Countries*, 6–31. Adelaide, Australia: University of Adelaide Press.
14. Brown, Lester R. 2004. *Outgrowing the Earth: The Food Security Challenge in an Age of Falling Water Tables and Rising Temperatures.* New York, NY: W. W. Norton.
15. Josling, Tim, David Orden, and Donna Roberts. 2004. *Food Regulation and Trade: Toward a Safe and Open Global System.* Washington, DC: Institute for International Economics.
16. Brown, Lester R. 2004. *Outgrowing the Earth: The Food Security Challenge in an Age of Falling Water Tables and Rising Temperatures.* New York, NY: W. W. Norton.
17. Verba, Sidney, Norman H. Nye and Jae-On Kim. 1971. *The Modes of Democratic Participation: A Cross National Comparison.* Beverly Hills, CA: Sage Publications.
18. Dittmer, Lowell, Haruhiro Fukui and Peter N. S. Lee (eds.). 2000. *Informal Politics in East Asia.* Cambridge: Cambridge University Press.

19. Brown, Lester R. 2004. *Outgrowing the Earth: The Food Security Challenge in an Age of Falling Water Tables and Rising Temperatures.* New York, NY: W. W. Norton.
20. Sansom, Goerge. 1963. *A History of Japan, 1615–1867.* Stanford, CA: Stanford University Press.
21. Ibid.
22. Garon, Sheldon. 1997. *Molding Japanese Minds: The State in Everyday Life.* Princeton: Princeton University Press.
23. Lewis, Michael. 1990. *Rioters and Citizens: Mass Protest in Imperial Japan.* Berkeley: University of California Press.
24. Ibid.
25. IPR 1933, 1.
26. Ibid.
27. Ibid.
28. Francks, Penelope. 2003. "Rice for the Masses: Food Policy and the Adoption of Imperial Self-Sufficiency in Early Twentieth-Century Japan". *Japan Forum* 15 (1): 125–146.
29. Hayami, Yujiro. 1972. "Rice Policy in Japan's Economic Development". *American Journal of Agricultural Economics* (February): 24–31.
30. Fisher, Galen M. 1937. "The Landlord-Peasant Struggle in Japan". *Far Eastern Survey* 6 (18) September: 201–206.
31. Ibid.
32. Hayami, Yujiro. 1972. "Rice Policy in Japan's Economic Development". *American Journal of Agricultural Economics* (February): 24–31.
33. Ibid.
34. Yoshiaki, Nishida. 2003. "Dimensions of Change in Twentieth-Century Rural Japan". In *Farmers and Village Life in Twentieth-Century Japan*, ed. Ann Waswo and Nishida Yoshiaki, 38–59. London: RoutledgeCurzon.
35. Ibid.
36. Aurelia George, and Eric Saxon. 1986. "The Politics of Agricultural Protection in Japan". In *The Political Economy of Agricultural Protection: East Asia in International Perspective*, ed. Kim Anderson and Yujiro Hayami, 102. Sydney: Allen & Unwin.
37. Francks (2003), Jussaume (2003), and George (2001).
38. Jussaume, Raymond A. Jr. 2003. "Part-Time Farming and the Structure of Agriculture in Postwar Japan". In *Farmers and Village Life in Twentieth-Century Japan*, ed. Ann Waswo and Nishida Yoshiaki, 199–219. London: RoutledgeCurzon.
39. Yoshiaki (2003), Waswo (2003), and Takashi (2012).
40. Yoshiaki, Nishida. 2003. "Dimensions of Change in Twentieth-Century Rural Japan". In *Farmers and Village Life in Twentieth-Century Japan*, ed. Ann Waswo and Nishida Yoshiaki, 38–59. London: RoutledgeCurzon.
41. Ibid.

CHAPTER 2

The Political Landscape: Recent Agricultural Policies and Rice Growers

The current policies affecting rice growers and their families have been created in the context of large-scale changes occurring throughout Japanese society. These include, but are not limited to the following: responses to the policies and actions of intergovernmental organizations such as the World Trade Organization (WTO), the Association of Southeast Asian Nations (ASEAN), and regional efforts to balance bilateral and global trade agreements; Japan's relationship with the United States and China; changing demographic trends, the most pressing of which is Japan's aging population and decreasing birth rate as well as crisis management such as the Fukushima-Daiichi catastrophe in the Tohoku region. Former studies[1] of Japan's policymaking bodies argued that Japan's diet, even when not under control of the Liberal Democratic Party (LDP) must balance sometimes extremely unevenly weighted priorities. In the arena of rice growing, the LDP has had a long-standing relationship with rice growers which one could argue, began soon after the Occupation and Supreme Commander of Allied Powers (SCAP, Douglas MacArthur) Constitution was adopted. The rural, mostly conservative majority of rice growers are LDP loyal, guaranteeing LDP success in those few times when the party has strong challengers.

However, with the last several election cycles and with the demise of rural farming affecting this voting block, rural interests and those of rice growers are less important to the LDP than they once were. Recently, under Prime Minister Abe, the LDP has shifted priorities with regard to

© The Author(s) 2019
N. L. Freiner, *Rice and Agricultural Policies in Japan*,
https://doi.org/10.1007/978-3-319-91430-5_2

agricultural policy because of the set of forces listed above that have challenged the government's ability to respond to external challenges while balancing the priorities of its domestic constituencies. Japan has yet to fully recover from the economic downturn of 2009–2010, although Abe's "three arrows of reform" are intended to address these concerns, the results they have achieved to date have been modest. With the high number of demands on its budget, the LDP has difficult choices to make in the coming years as Japan's aging population will grow to one-third of the total population as well.

POLICYMAKING IN THE JAPANESE NATION-STATE

In Japan, the agricultural policy framework discussed in this book focuses on public institutions, private actors, laws and policies as well as informal social currencies, including nationalism and the role of cultural performance in examining policies with regard to rice farming. The primary role of the bureaucracy in legislation underscores the centralized nature of the Japanese state. Although attempts have been made to strengthen local initiatives, policy originates within the bureaucracy and is closed to public scrutiny and debate. For agricultural policy, the Ministry of Agriculture, Fisheries and Forestry has (MAFF) until recently been the primary creator of agricultural policy, working with the Ministry of Environment when their tasks overlap (when regulating genetically modified organisms, for example). Along with these centralized institutions within the Japanese government, an informal group of policymakers called *norinzoku* (for agricultural tribe) in the Liberal Democratic Party are discussed as well as JA Zenchu, the cooperative responsible for distributing government funding and implementing policy. These three bodies taken together are the primary institutions responsible for rice policy, including both its creation and implementation. Along with these structures at the central or national level of politics, this book also discusses international trade policy, and the voices of farmers, those affected by this policymaking machinery as well as those from consumer groups motivated by food safety issues. The Japanese bureaucracy plays two roles in legislation: policymaking and policy implementation. Unlike many other advanced democracies, the policy process begins in the bureaucracy in Japan with most laws being formulated by government bureaucrats. The practice of *gyousei shidou* 行政指導 (administrative guidance) is also exercised by bureaucrats who may issue administrative ordinances that have the force of law or communicate informally to persuade social groups to comply with a particular policy. As noted earlier in the Literature Review, previous researchers have at times advocated the high degree of centrality and coordination of Japanese bureaucracy. The political process operates out of public view in large measure, and decisions are made within the context of bureaucratic authority, among officials who have close personal ties and a high degree of mutual trust. However, this does not mean that the political process is a consensual one. In fact, Kawanaka[2] uses the term "jungle warfare" to portray the prevalence and ferocity of intra-bureaucratic conflict in his analysis of administrative bureaucracy in Japan. This study argues that while the bureaucracies have access to a legal framework of intervention that is highly developed, there are a complex set of pressures including international commitments and domestic actors to which they are responding. Moreover, the degree of conflict even within individual ministries is high and overall bureaucratic policies illustrate conflicted interests and a lack of cohesiveness and coordination rather than uniform policies that have a highly coordinated set of aims.

In elections, during the past five years, there was considerable debate about the evolution of a two-party system in Japan as the LDP lost a clear majority and the Democratic Party of Japan (DPJ) made enough gains to contend the dominance of the LDP. In the 2004 election, there was a clear two-way battle in nearly every constituency and polls showed the two parties neck and neck.[3] In the short term, this apparent reversal of fortunes seemed insubstantial because the LDP gained supremacy in the 2005 election under the leadership of Junichiro Koizumi. In the 2009 House of Representatives elections, the DPJ called on voters to support a shift in political power to change the course of the country amidst the global economic crisis and criticism of Prime Minister Aso, the DPJ won the largest force in the House of Representatives for the first time in history. This victory marked a turning point in Japanese politics as the LDP performed the worst since the party's inception, winning its worst share of seats (119 of 480). The turning point was short lived though as in the next election season, the LDP once again asserted its dominance. The LDP that was responsible for recreating and industrializing Japan after WWII has controlled the political landscape for much of the last 70 years. Although the Emperor of Japan no longer posses formal political power, he is the Head of State, which emphasizes the conservative social order that prevails in Japan. In recent years, the Prime Minister (Head of Government) has consolidated the governmental ministries, strengthening the government.

Although there were challenges to LDP dominance in the late 1990s and in 2005, these challenges have not altered the party's dominance. Even a set of electoral restructuring in 1994 which introduced proportional representation in some districts and abolishing the single nontransferable vote (SNTV) system. The reform ended the system which privileged rural voters, and Japan's party system underwent a number of recombinations providing voters with more alternatives. Despite these changes, however, Japan's LDP rulers held onto power despite numerous attempts to replace them.[4] The national level agricultural policy framework of Japan's LDP, the norinzoku and the MAFF continue to wield significant power. This is true despite a number of changes that were predicted to bring meaningful reform and to break up the agricultural policy regime which has persisted for over 50 years. First, changes to the electoral system in 1994 took place which replaced many multi-member districts with single member ones, encouraging competition a more balanced shift of representation between rural and urban areas. Previously, rural areas enjoyed a higher allocation of votes than was fair based on population, because the multi-member districts (which a candidate can secure by winning a lower proportion of the total vote than single member districts which require over 50% of the vote to win) permitted special interests to be represented disproportionately. Individual candidates now must appeal to a broader range of voters so diet members who specialize in agriculture must cater to a variety of interests rather than advocating just for agricultural interests. The electoral system changes also affects farm voters, who are now less sensitive to the appeals of cooperatives who at one time who able to mobilize support for political candidates, allowing them substantial ability elect politicians that would advocate for cooperative interests. Secondly, consolidation at the Prime Minister's level that began under Koizumi, means increased scrutiny and oversight over policy areas like agriculture. PM Koizumi consolidated the number of ministries and created an apparatus for policymaking that reaches across all policy areas. PM Abe increased ministerial power by creating oversight bodies that direct policy, the primary ones are the Council on Economic and Fiscal Policy (CEFP) and the Council for Regulatory Reform (CRF). These two advisory councils at the executive level oversee agricultural issues and policymaking, they have the power to direct ministries such as the MAFF; their members are ministers of state, and the structures have the task of producing policy proposals for specific issues and providing oversight on fiscal policy among others. These structures alter the previous model of LDP policymaking in which LDP politicians prepared budgets and formulated policy with little oversight or direction from above. Now, policy is more of a top down process, and policy areas have far less freedom in creating policy and making budget requests on their own. They must heed the directives of executive level bodies and the goals of a more powerful and involved executive. Despite these changes, the agricultural sector continues to intervene in agriculture and to weather both the pressures from the executive level and electoral reform because it serves both the institutional and private interests of Ministry officials and JA Zenchu who work closely together. As Mulgan notes, executive pressure has hardened the interests of LDP members who are norinzoku to illustrate unity and to continue to defend agricultural protection and income support programs for farmers.[5] During the most recent election for the lower House of Representatives (*shugiin* 主義員) the LDP campaigned on a platform of economic revival, pulling Japan out of its recession and maintaining the commitment to using nuclear power. The opposition DPJ campaigned on the nuclear reactor issue as well but said that it would close all reactors by the 2030s. The LDP won a landslide majority with 294 out of 480 seats regaining power after a mere three years as the opposition, the DPJ went from having 308 seats to 57. Amidst the Fukushima-Daiichi meltdown catastrophe and ensuing crisis the Japanese returned power to the traditional and safe LDP. Since 2010, the Japanese economy has been at a standstill, the tsunami on March 11, 2011 caused supply chain disruptions, power supply restrictions, and a nuclear meltdown, creating an economic slump and a bad fiscal position. The return of the LDP to power is most certainly linked to Japan's economic uncertainty, although recently the economy has shown some signs of recovery, it is still very weak.

In Fall of 2017, Japan's population once again reaffirmed Prime Minister Shinzo Abe's leadership with a sweeping victory for his LDP in Japan's House of Representatives. The surprising call for the snap election in the House came in late September, after North Korea had fired another missile, with its longest delivery system yet on September 15 and a rise in approval ratings in a recent set of opinion polls seemed to provide Abe and his party, the LDP reason to gamble on Japanese citizens history of voting for the LDP amidst crisis by calling new elections. The gamble paid off and with the LDPs victory in place, Abe is likely to become Japan's longest serving Prime Minister. Furthermore, with the renewal of Abe's party continued dominance, plans to promote his nationalist, conservative agenda are likely within the bureaucracy as well. For agricultural policy, this means moving forward with reforms, including the ones targeted at JA Zenchu, advocating the washoku diet, promoting this diet internationally to increase Japan's exports and proceeding with structural changes to the farm sector, including farmland consolidation which will be discussed below. The degree to which these changes will erode the agricultural policy framework is, as yet, arguable.

The Ministry of Agriculture, Fisheries and Forestry

Within the Japanese government, the national administrative framework responsible for creating and implementing policy affecting rice farmers originates in at two primary policymaking bodies: the Ministry of Agriculture, Fishers and Forestry (MAFF), or 農林水産省 *Nourinsuisanshou* in Japanese and the Ministry for Economy, Trade and Industry (METI) (経済3号証, *Keizaisangyoushou* in Japanese). Taking a long view of the history of policymaking of the two ministries illustrates a dynamic of conflict, with either body winning at certain times depending on the state of the Japanese economy and how power is distributed in the LDP. From the 1990s until today, the METI has lost power with the economic turndown in Japan, the low birthrate, and international pressure. This argument is less applicable to the MAFF for reasons which will be outlined below. Both ministries are subject to changes such as Cabinet reshuffling and reorganization that originate with the Prime Minister. The most concerted and dramatic of these occurred in 2001, when Prime Minister Koizumi consolidated the Ministries down from 24 ministries and agencies into 11 ministries (4 of these are super sized) and 2 agencies that report directly to the Cabinet Office. This reform was in part a response to corruption in the bureaucracy and mistrust of bureaucrats by the general public. As a result of this change, government (Prime Minister and Cabinet) was strengthened and the executive branch has enhanced powers. The Cabinet has a larger Secretariat, an enlarged Cabinet Office (now led by the Prime Minister) and bodies such as the Council on Economic and Fiscal Policy attached to it that makes budget decisions (it was once part of the Ministry of Finance). In the reorganization of 2001, the ministries that are pivotal for rice policy and also the focus of this book experienced some internal restructuring but maintained their administrative and jurisdictional boundaries and both continue to hold significant power in their policy fields that were not diminished with the restructuring.

The MAFF is the central administrative policy body responsible for agricultural policy. It is subdivided into individual bureaus and also agencies. The bureaus subdivide the ministries and in some cases also have regulatory and implementation powers. As was the case with the Ministry for International Trade and Industry or MITI (the former title of METI), the ministries relationship with the industries that they represent is often one that is cooperative (such as the one detailed by Chalmers Johnson[5] in his pathbreaking work) rather than divisive. These cooperative relationships mean that policymaking bodies like the MAFF often include agencies that work directly with those whose interests to which they are most responsive. In the case of the MAFF, this includes the agricultural cooperatives, such as JA Zenchu and JA ZEN-Noh, which are administrative support organizations (行政補助期間, *gyousei hojo kikan*) that are essentially extensions of the MAFF itself.

One of the key methods of intervention up until 2015 was through the Food Control system which governed the collection, storage, and distribution of rice, thereby also acting as a price control mechanism. The system was not intended to favor rice growers, as Mulgan argues, but to maintain a system of intervention itself which preserved the network of control that the Ministry had established for itself. As Mulgan observes,

> Given the prominence of rice farming in Japanese agriculture, intervening in all aspects of the rice market meant virtually controlling Japanese agriculture. The MAFF followed the principle of rice market supremacy, *kome shijou shugi* (米市 場主義). Throughout the five or more decades of the FC system's operation, the MAFF sought to maximise rice market intervention by drawing out the process of reform and initially only permitting adjustments tat the margins of the rice distribution control.[6]

As noted earlier, price supports for farmers have gradually waned, the rice production adjustment or *gentan*, began in the 1970s and though it has faced challenges from other agencies in the bureaucracy as well as large-scale rice producers, many authors argue that the price control system remains a stalwart of the MAFF. The New Food Law of 1995 had the potential to challenge the *gentan* and to finally upset the MAFF's control of the rice market. The new law changed regulations concerning rice collection, wholesale, and retail distribution. The law recognized three channels for rice marketing, *seifumai, jishu ryuutsuumai,* and *keikakukugai ryutsumai.* The first two categories were synonymous with existing categories but were recreated as a new category of "orderly marketed rice." The law established the Voluntary Marketed Rice Price Formation Centre, that would operate as a wholesale market for *jishu ritsumai* with prices from wholesalers acting as an index both for the cooperatives and the local market. However, bidding at the center was regulated which altered its ability to reflect actual supply and demand and the pricing that would concur with both. A study by the Department of Foreign Affairs noted that

> The relatively high level of administered prices evident in 1996 together with the restriction on price fluctuations in the Voluntarily Marketed Rice Price Formation Centre, combine to direct upward pressure on prices of about one third of domestically produced rice, which will indirectly influence the rest of the market. The end result is that rice prices in Japan are likely to remain at artificially high levels.[7]

Prices continued to be higher for consumers while the MAFF retained control over the rice market and insulated it from the market. The third channel for rice growers was for "non-orderly marketed rice" and allowed growers to bypass the other two routes to sell directly to wholesalers, retailers, and consumers. Godo[8] comments, that the opening up of this third channel was merely legitimization of a situation that already existed in Japan, the nonlegal trade in rice, or black market. Making this channel legal did not expand the number of growers who used it. Despite its lofty goals of liberalization, in truth, the Ministry and its Food Agency were more interested in preserving their own administrative control and supervision in order for it to continue to exist. Rice grown in Japan continued to be heavily administered under the two most powerful functions of the MAFF's bureaucracy: the administrative distribution control system or *gentan*, and the restriction of market access for foreign rice. These policies have also had long-term and deep effects on the global market of rice, which is more volatile than any other global grain. Government policies such as the storage of rice and the *gentan* have made the market less predictable, creating unnecessary volatility a topic that will be addressed at greater length in Chapter 3.

This control had a variety of effects but the groups who are most affected are consumers and rice growers. Without the removal of the *gentan*, Japanese people would continue to pay higher prices for rice. Moreover, rice growers exist in a system that is unable to respond to the market and to modernize the way that other agricultural producers around the world already have. The long-term effects of these controls play an important role in shaping rice growing in Japan and may in fact be one of the reasons that rice growing in the future is a field less and less likely for young Japanese to enter. Eventually, rice growers in Japan will die out, leaving the country with a stunted, nonreactive and aged system of production and distribution with the Ministry that is supposed to represent their interests at fault.

POLICY OVERVIEW

Prior to the 1990s Japan had a two-tiered pricing system that caused the production of excess rice, forcing the government to introduce an acreage control program in 1995. This two-tiered system was a product of the Ministry of Agriculture and Commerce (MAC), a body with divided interests. The division between those responsible for agricultural policy with those responsible for industry and trade policy led to intra-bureau conflict and incoherent policymaking. The Agricultural Minister held a conservative view that small-scale farming was the backbone of Japan's social system with rural farmers preserving traditional values[9] (Takahashi 2012). However, those responsible for trade and industrial policy argued for more efficient, mechanized agriculture that requires larger fields. In these inter-agency battles, the Agricultural Minister (a member of a powerful LDP cohort) won for the most part. The government passed two laws that subsidized small rural farmers in the early 1990s, the Food Control Law and the Staple Food Law. These laws introduced a compulsory system of selling rice to the government. Takahashi argues that three policy measures affected rice prices at this time: (1) Government purchase of rice, (2) Output payment (such as an income stabilization program for rice farmers), and (3) Acreage control. The acreage control program or *gentan* diverted a portion of rice paddy fields to developing other crops such as wheat and soybeans. Acreage control is a decades-old policy first implemented to assist with farmer's incomes following WWII. The policy was intended to decrease the supply of rice, thereby increasing the price. Arguably, these policies allowed Japanese rice farming to remain insulated from the pressure to modernize. However, soon Japan would be forced to reconsider these policies in the wake of its participation in global trade agreements and ministerial changes that shifted power.

Japan's policy shifts in the past 20 years present one with a story of adjustment rather than conforming or adapting to international agreements, specifically the Uruguay Round Agricultural Agreement (URAA), which in 1995, ascended to the WTO Agreement on Agriculture (AOA). Both the URAA and the WTO AOA obligate countries to open their economies to global trade by limiting import tariffs and import restrictions and eliminating subsidies and other supports for domestic producers. The government agency responsible for formulating Japan's response to these agreements is the MAFF. Unlike its predecessor the MAC (Ministry of Agriculture and Commerce which existed prior to WWII), the MAFF is singly responsible for advocating on behalf of the interests of Japan's farmers and fisherman with environmental protection, land, and forest management as additional policy drivers. The Ministry's own webpage states "Agriculture, forestry and fisheries industries, as an important sector of Japan's economic structure, contribute outstandingly to the development of national economy and stabilization of national life through their role of providing stable supply of foods indispensable to our daily life".[10]

Mulgan and other scholars[11] argue that in the late 1990s and early 2000s the MAFF faced a restrictive policy environment. International pressures called for the elimination of subsidies and protections for farmers while internally, structural reform that included ministerial consolidation was beginning under Prime Minister Koizumi. The structural reforms were undertaken as part of a process of economic reform, which called for shifting resources to more productive industries and promoting greater competition. While farm policy was not the focus of these reforms it was influenced by the goals and changes in organization that was part of the process. The restrictions on land use which had long been advocated for by the MAFF was viewed by policymakers in competing ministries (such as METI) as an obstacle to changes in land management that would create larger, more efficient farms.

The MAFF responded to these dual pressures in three ways. First, the Ministry used early rice tariffication (in 1999), in advance of the 2000 deadline established by the WTO. Second, it passed the New Basic Law in 2003 (also known as the Revised Food Law). Third, it created the Rice Policy Reform Charter of 2002 followed by the Law for Stabilization of Supply, Demand and Price of Staple Food. As a group, these new policies allowed the MAFF to continue to protect rice farmers even while acceding to WTO policies and maintaining Japan's commitments to international agreements.

EARLY TARIFFICATION

Under the rules established by the WTO during the Uruguay Round in 1994, member countries were obliged to open trade to competition and to weaken measures that would protect domestic industries. At this time (1999), the Japanese rice market was already oversupplied by domestic producers. The government was buying back rice surpluses and paying farmers not to grow rice in favor of other crops. This made decision-making in the Ministry very difficult. How could Japan let in more foreign rice while domestic producers were oversupplying it? The Ministry's answer was to introduce a tariff on rice in the final year of the URAA in advance of stricter obligations and use the calculations of the new agreement in Japan's favor. Increasing the rice tariff early would actually allow Japan to reduce Japan's rice import obligations, from .8 to .4% each year over six years.[12] Introducing the rice tax early allowed Japan to better its negotiating position because it was playing by the rules and would no longer be seen as using rice as an exception to the new trade rules. Japan used the rules of the agreement in its favor to continue to protect its domestic rice market.[13]

NEW BASIC LAW

Domestically, the new law on food reframed the MAFF's approach in a number of policy areas; I argue that the most important aspect of the new law is the linking of food with sustainability and energy conservation issues. Mulgan argues that the law reflects the importance that the Ministry attached to gaining public support for its policies thereby upholding the legitimacy of its expensive protectionist regime.[14] Certainly the law is written with a number of public interests in mind, and specifically it addresses consumer groups and environment and sustainability-focused groups. Both consumer groups and environmental groups in Japan are substantial in number and have the power to influence policy, with a high percentage of Japanese supporting one or both causes.[15] The first article of the law addresses international agreements and clearly situates food policy as a public safety issue,

> In consideration of the vital importance of precise responses to the development of science and technology, and to the progress of **internationalization** and other changes in the environment surrounding Japan's dietary

habits, the purpose of this Law is to comprehensively promote policies to ensure food safety by establishing basic principles, by clarifying the responsibilities of the state, local governments, and food-related business operators and the roles of consumers, and **establishing a basic direction for policy formulation, in order to ensure food safety**. (my emphasis)[16]

With food safety now expressed in Japanese law as the responsibility of the state and state agencies, Japan could use this instrument in international negotiations to strengthen its position when necessary. In legal terms, the law made food security a nonnegotiable prerogative of Japanese trade posture (Chart 2.1).[17]

The New Basic Law (Law No. 127) amends and replaces the Basic Food Law of 1961. The Law contains a number of provisions which strengthen the prerogative of the MAFF and the negotiating position

Chart 2.1 Outline of New Basic Law on Food, Agriculture and Rural Areas

Chap. 1: General Provisions	Priority on securing quality, reasonably price stable food supply
	Reliance on domestic supply as basis
	Role of agriculture in conservation, water, and maintenance of cultural traditions
	Focus on development of rural areas and role of farmers, farmer's organizations
	Outlines responsibilities and duties of state and local governments, legislature, consumers, and the food industry
Chap. 2: Basic Principles	
Section 1	Food self-sufficiency as a key goal
Section 2	Use of tariffs and import restrictions when domestic industry threatened
Section 3	Securing farmland for agriculture and farm management
Section 4	Support of rural areas to compensate for disadvantageous conditions
Chap. 3: Administrative Bodies and Relevant Organizations	Reorganization and restruction of administrative organs
Chap. 4: Council of Food, Agriculture, and Rural Area Policies	Creates the Council, members are appointed by the Prime Minister

of the Japanese government with regard to international commitments because of the linkage of food safety with security. Throughout the law, security is mentioned a number of times. In Introduction of the law, the priority of "securing a stable food supply" is listed as one of the primary reasons for creating the law, along with a focus on the development of rural areas, which are viewed in the law as repositories of cultural tradition requiring special attention. The target of the law outlined above is described in detail in Chapter 1 of the law, where Basic Principles are discussed. In each of these sections, the key goals of the law are listed. They are: (1) Food self-sufficiency, (2) Control of imports to protect domestic producers, (3) The ability to adjust prices, and (4) Support to rural areas. The law is in essence the codification of powers of the MAFF, in order to preserve the Ministry's role in agricultural trade. The powers outlined in the law at times expressly counter those contained in international agreements such as the WTO's Agreement on Agriculture, as well as the Trans-Pacific Partnership (TPP). For example, the control of imports is expressly forbidden by both agreements, except when certain industries are threatened.

Linking food security to sovereignty also elevates the importance of food and the Food Agency within the MAFF. The adoption of the New Basic Law guaranteed the role of the Food Agency and solidified its power.

RICE POLICY REFORM CHARTER

More than any other measure, the Rice Policy Reform Charter (or Outline) will reshape Japanese rice farming in drastic ways. The "Outline of Rice Policy Reform" was established in December 2002, based on which the Basic Plan for Rice Policy Reform was drawn up in July 2003. At the same time, the Law for Stabilization of Supply, Demand and Price of Staple Food was revised in its entirety. The main points of rice policy reform are as follows: (1) Producer's groups should take the initiative in rice production adjustment, (2) The planned distribution system should be abolished, and all the rice on the market should be freely distributed, and (3) A set of farm management measures intended to stabilize the supply of rice and other crops targeting large-scale paddy field farmers and agricultural corporations should be implemented. The reform established a fund of contributions from producers in addition to a government subsidy that provides an income supplement in the event of income loss.

The producer's groups referred to are the *nokyo*, Japan's agricultural cooperatives; the largest of these is JA Zenchu. JA Zenchu is a powerful farm lobby that has traditionally held significant power in the LDP that has ruled Japan since WWII except for two brief periods of opposition rule. The elimination of the plan distribution system refers to the government buyback of surplus rice, a policy that has incurred significant cost. Without this program, rice will be allowed to sell at significantly lower market prices, abolishing a trade distortion, which kept rice prices high enough for small-scale farmers to earn a modest living. The farm management stabilization measures target large-scale paddy field farmers with farms over 4 hectares (10 in Hokkaido). The measure allows local governments to encourage large-scale production and gives local governments the power to use the subsidy (funded in part by the central government) how they wish. Farm management measures also encourage the creation of larger scale farming through the creation of agricultural corporations. The government envisions the corporations as a way for small-scale farmers to work together to create larger businesses thereby helping to preserve the legacy of small family paddy traditions. Over time, Japanese policymaking has responded reactively, rather than proactively to government participation in international agreements while attempting to maintain some control over rice subsides as Table 2.1.

Taken together these measures have significantly and detrimentally affected the ability of small paddy farmers to survive, to earn a living growing rice and to pass on that living to their children. There is some hope given that local governments have control of funding and may in fact help subsidize small farms who try new forms of entrepreneurial farming and those small farmers and families who can organize to form corporations may benefit as well. All of the farmers I spoke with are most

Table 2.1 Japan's response to international agreements

Policy measures on agriculture in Japan		International agreements	
1990s	Food Control Law	1995	WTO AOA
	Staple Food Law		
1999	Rice Tariffing	2000	URAA
2002	Rice Policy Reform Charter	2001–2008	Doha Round
2003	New Basic Law (Revised Food Law)	2015	TPP

concerned about the potential passage of the most recent international agreement affecting rice policy: the Trans-Pacific Partnership.

NORINZOKU, INTRA-BUREAU CONFLICT, AND MINISTERIAL CHANGES

One of the key methods of insulating agriculture from international pressure and reform from above has been through *norinzoku* loyal administrators maintaining leadership of the MAFF and acting as gatekeepers to stave off the pressure of drastic policy reform. As Davis and Oh (2007)[18] argue, "MAFF policy autonomy allows the ministry, with the active support and cooperation of its partners, nokyo and norinzoku politicians, to reduce the impact of reforms through delayed targets, side payments, and other measures" (p. 23). Several noted authors, including Mulgan, Davis, and Calder detail the relationships among LDP politicians including norinzoku, the nokyo and the MAFF. Mulgan goes so far as to argue that this relationship constitutes an "iron triangle" a la Chalmers Johnson. In order for this iron triangle to maintain control over the agricultural policy arena, it must be able to coordinate across policy environments and policy levels (domestic and international). While the policymaking bodies constitute a network of interests, it is not as coordinated either as Mulgan or Davis and Oh suggest. The *norinzoku* itself has varying constituencies within itself with politicians and lobbies favoring different prefectures across Japan, a high number of agricultural products as well as differences in types of producers (those who favor small, low-intensity farming and those who view mechanization as the way forward). Moreover, the MAFF itself is known for probably the strongest intra-ministry sectionalism in the central government among its bureaus and even within the same bureau. It may be appealing to characterize Japanese bureaucratic politics as an iron triangle, but this characterization does not ring true. The policymaking process is far more complex, with varying interests among the *norinzoku* lobby and even within the ministry itself to argue such coordination applies to today's politics.

The final institution that makes up this supposed iron triangle is Japan's most important organization for farmers, the *nokyo*. As discussed earlier, the *nokyo* has pre-WWII history and after WWII the *nokyo* became one of the most dominant and compelling interests in Japanese politics. By far the largest of the nokyo is JA Zenchu, the Central Union of Agricultural Cooperatives. For the past 60 years, JA Zenchu has dominated Japanese rice politics but it has lost some power after being the subject of government reform in 2015 (it is too soon to tell how significant this loss will be). The reform deprives the lobby of its privileged status to oversee local cooperatives, which reformers hope will give greater freedom to local cooperative organizations by allowing them to control their own funds and to be audited publicly.

After Japan's most recent elections, Prime Minister Abe removed staff members loyal to the *norinzoku* or agricultural tribe in the LDP. The Ministry's were re-organized to emphasize the focus on promoting economic growth, at least that is what has been stated in public, one such change is clearly expressed in the changes to the MAFF. When the MAFF was reorganized in 1978 (formerly the MAC directed all agricultural and economic policy) the top Minister and most members of the MAFF were loyal members of the *norinzoku* cadre of the LDP. Slowly, however, these tribe members have been replaced. In October 2015, Hiroshi Moriyama was appointed the new Minister of MAFF, followed by Yuji Yamamoto in August 2016. Yamamoto was the former Chair for the Committee on Economy, Trade and Industry. Furthermore, the remainder of the Ministry's chief officials are mostly lawyers and loyal to the METI. They include State Minister of Agriculture, Fisheries and Forestry Ken Saito, an Economist educated at Tokyo University or *Todai* who is a long-time supporter and advocate for economic interests over agricultural ones even before MITI became METI; Parliamentary Vice-Ministers Katsuo Yakura, also a lawyer and with strong ties to METI and Kenichi Hosoda, with ties to METI and nuclear power issues. With the exception of two members of the Ministry with international expertise, all of the new leadership of the MAFF has either worked directly for METI or have strong ties to METI through other organizations (see Table 2.2).

It is clear from these recent shifts on the MAFF, taken together with the decreased power of nokyo as a result of changes to agricultural policies, that the former network that enabled farmers and their advocates to protect rice production from external competition and large-scale mechanization is beginning to break down, albeit the change is likely to be very slow. Though the full effect of some of these efforts won't be felt until the laws come into force in 2019, the impact of some may come sooner. In Fall of 2018, policymakers will take up agricultural policy reform wholesale and with the MAFF now totally in the hands of economic policy wonks, it is likely that they will take up drastic measures to change Japanese rice growing.

Table 2.2 Current leadership in Japan's MAFF

Minister of AFF	Yuji Yamamoto	Chair of budget Cmte, Former Chair Cmte on Econ, trade and industry
State Minister of AFF	Yosuke Isozaki	Special Advisor to PM, International Affairs Office
State Minister of AFF	Ken Saito	MITI
Parl Vice-Minister	Kenichi Hosoda	METI, Nuclear Power Issues
Parl Vice-Minister	Katsuo Yakura	METI
Vice Minister	Masaaki Okuhara	Management Improv. Bureau MAFF
Vice Minister for IA	Hiromichi Matsushima	Exp with UN, Int'l org's, WTO, with MAFF since 1982

LANDOWNERSHIP AND CONVERSION

The Agricultural Land Law (ALL) of 1952 states that its purpose is to "stabilize the status of cultivators and boos domestic agricultural production" by regulating land conversion to nonagricultural uses and promoting the rights to land by farmers (ALL, January 8, 2017).[19] The law provides that land used for agriculture must be owned by full-time farmers or cooperatives and that those lands are used for agricultural purposes only. The law also details the restrictions on the transfer of land and stipulates that agricultural land should continue to be used for either cultivation or livestock farming unless permission is granted according to a set of conditions outlined in the law.

As mentioned earlier, in Introduction, the SCAP completed dramatic land reform in Japan following WWII which was successful in transforming former feudal-like landlord–tenant relationships so that many Japanese became landowners. After WWII the Japanese government enacted two complimentary pieces of legislation which established the framework for agricultural policy and landownership. These two laws are both based upon the farmer owner principle, to prevent the landlord system from reasserting itself and to protected farmers by strictly regulating the ownership, size, use, and renting of land.

Following up on land reform Japan enacted the Basic Law on Food Agriculture and Rural Areas of 1949, which outlines the basic government policy areas regarding food, agriculture, food self-sufficiency, and establishes areas of government support. Articles 21 and 23 of the Basic Plan privilege full-time family farms as necessary for revitalizing agriculture "the State shall take necessary measures for revitalizing family farming by means of bringing about conditions for better farmers' management"[20] (Basic Law, p. 6).

Taken together the two laws form the basis for agricultural landownership and support for such ownership by the federal government, the ALL also established the Agricultural Committees, local bodies which oversee and implement the law by regulating the lease and sale of farmland. In order to rent, lease or sell farmland farmers must first seek permission from the local Agricultural Committee upon its approval, the request is then submitted to the prefectural governor.

On paper, the legislative prescriptions look highly restrictive in terms of converting farmland to nonfarm uses. However, in practice farmland conversion can be highly profitable as Godo (2001)[21] argues, landowners who are able to manipulate the land use regulations can convert land, cashing in on high prices by selling to developers, construction companies, or other private bodies. The regulation of farmland conversion is complicated and regulated by a number of prescriptions that must be met in order to legally convert land to nonfarm uses (see Table 2.3).

As is the case with many democracies, the use of farmland in Japan is subject to a number of protections and regulations. The Law Concerning Construction of Agricultural Promotion Areas at the national level exists alongside municipal zoning. Municipal governments are authorized to designate Exclusively Agricultural Areas or CAAs. Within Exclusively Agricultural Areas (EAAs), land is only to be used for farming and conversion of land to nonagricultural uses is prohibited in return for receiving subsidies (the MAFF also has set aside money for investment in these zones). The Agricultural Land Law and

Table 2.3 Laws regulating farmland use and conversion

	Adopted
City/Urban Planning Act	1968
Law Concerning Improvement of Agricultural Promotion Areas	1969
Renamed Agricultural Management Framework Reinforcement Act	1993
Agricultural Land Law	1952
Land Improvement Act	1949
Act on the Establishment of Agricultural Promotion Regions	1969
Community Areas Development Act	1987
Act on the Promotion of Community Farms	1990

City Planning Law (CPL) are also key mechanisms for managing land use. The ALL regulates farmland conversions, requiring the prefectural governor's permission when a farmer wants to convert land to nonagricultural uses. To obtain such permission, the farmer must submit a plan for conversion and explanation for why he wants to convert. Once he submits his plan and explanation, the local Agricultural Committee classifies the farmland into four types (Types A, 1, 2, and 3) according to the quality of land and potential affect on neighboring farmers. The conversion of larger paddy fields (over 4 hectares) is governed by the MAFF (Godo 2007).[22]

Laws related to urban planning, such as the CPL, also regulate farmland use and conversion. The CPL regards the conversion of farmland as land development. All land development in Urban Control Areas (UCAs) requires permission from the local Development Committee. For farmers to convert land that is included in both an EAA and a UCA the farmland must first be excluded from the EAA (requiring a municipal governor's revision of the EAA), then permission is needed for farmland conversion from the ALL and finally the farmer (or land developer) needs permission from the CPL. Few researchers have attempted to tackle the issue of the manipulation of farmland use regulations by farmers.

Along with the ALL and the Basic Act, the City Planning Act was established in 1968. This law regulates city planning, especially the transition zones on the outskirts of cities. According to the law, urban areas and neighborhoods were classified into areas where urbanization was promoted (urbanization promotion area) and areas where urbanization was controlled (urbanization control area). The City Planning Act also implicated agricultural land on the outskirts of cities, oftentimes as part of an urbanization control area. In order to specify and preserve farmland the MAFF enacted its own legislation to protect its interests and Hirasawa argues in opposition to the City Planning Act. The MAFF law is called the Act for Improvement of Agricultural Promotion Areas. The Act specified agricultural lands and surrounding areas as areas for agricultural use and prohibited the conversion of such lands in the areas for agricultural use.

The Law Concerning the Improvement of Agricultural Promotion Areas, which was renamed the Agricultural Management Framework Reinforcement Act (AMFRA) gives authority to regional and city governments to zone EAAs for the promotion of agriculture according to the Basic Law. Farmers in these specially designated areas are responsible for using farmland only for the purpose of farming and these farmers receive priority when the MAFF allocates subsidies. Farmers in these zones are strictly regulated in their ability to convert land to nonfarm uses, most farmland, including rice paddies (80%) are in one of these zones. Paddy fields used for rice growing are regulated by the ALL and when a farmer wants to convert the rice paddy for nonfarm uses, permission is required from the local Agricultural Committee as well as the prefectural governor.

A number of new laws and revisions to existing laws over the years included those listed above have attempted to reform Japanese land management, particularly rice growing, by encouraging the conversion of small paddy fields into larger ones. The 1969 Act Establishing Agricultural Promotion Regions was revised in 1980 to promote the lease of large-scale agricultural operations by allowing ownership groups to be groups of individual farmers, and in 2000 when the act was renamed it allowed stock companies (those with nontransferable stocks) to create agricultural production corporations in order to lease land.

Major revisions to agricultural policy with regard to landownership and rice paddy field conversion took place when the ALL was amended in 1970. Control over renting land was substantially liberalized, and the owner farmer principle was converted to the cultivator principle which allowed tenant farming for the first time since WWII. These measures, which were intended to create larger paddy fields by way of consolidation, did not advance substantially with this change, however. The failures of this legislative change led to the creation of a new law intended to channel funding from the MAFF to farmers (again with the stated goal of paddy field conversion), called the agricultural land use promotion project established in 1975, and the expansion of the project was promoted with the Agricultural Land Use Promotion Act in 1980, and the (renamed) AMFRA in 1993. Its basic concept is that controls, such as permission for renting, specified in the Agricultural Land Act are exempted if certain conditions are satisfied. At the same time, it also aimed not only to promote liquidation of agricultural land but also to concentrate agricultural lands in the hands of principal farmers in the area through community discussion.

Another set of revisions occurred to the ALL first in 2003 and again in 2009. In 2003, the MAFF launched a new program whereby the authority that gives permission for farmland conversion would be transferred from the prefectural governors to the Agricultural Committees, eliminating a step in the existing process. Previously, agricultural committees would come to a decision which was recommended to the prefectural governor who would give final permission. The MAFF introduced this program under the pretext of decentralization. Since the members of the Agricultural Committees are farmers, Godo argues that this decentralization is likely to promote self-serving farmland conversion. Another noticeable shift by the MAFF is the provision of more chances for nonagricultural companies to purchase farmland.

According to the ALL, the ownership of farmland is allowed only for farm households and a special type of private corporation. These special corporations are referred to as "agricultural production legal entities." When a nonagricultural company wants to purchase farmland, the company starts up an agricultural production legal entity as an affiliated company. The MAFF is moving toward relaxation of the requirements for the qualification of agricultural production legal entities. In 2000, the MAFF allowed stock company-style agricultural production legal entities to be set up—before 2000, only limited liability company-style companies were allowed. In 2003, the MAFF also relaxed the qualification requirements for investors in agricultural production legal entities. These deregulations make it easier for nonagricultural companies to own farmland (See Table 2.4, below).

In 2009 the MAFF began a new reform package to the ALL to address the continuing weaknesses in existing law and to address the problem of slow conversion to larger paddy fields and to also begin the process of bringing idle or abandoned farmland into a system where it could be better managed. The revision encourages more nonfarming entities to start or participate in farming operations by relaxing regulations on farmland sales and leasing. The revision aimed to increase

Table 2.4 Major Revisions to Land Laws

Legal measure	Adopted	Revised
Agricultural Basic Law	1961	Every 5 years
Law Concerning Improvement of Agricultural Promotion Areas	1969	1980, 1975
Renamed Agricultural Management Framework Reinforcement Act	1993	1999
Agricultural Land Law	1952	1962, 1970, 2000, 2003, 2009

production by shifting the focus of its legislation from "possession" of agricultural land to "use." The government hoped to accomplish this shift by allowing nonagricultural companies to lease farmland without going through local governments, a fairly drastic change as this was the first time nonagricultural businesses were allowed to use farmland. According to the revision, nonagricultural companies were allowed to increase their share in agricultural corporations from 10 to 25%. Also, private companies who partnered with farmers, and were officially recognized by MAFF as part of the government's agricultural revitalization efforts, were allowed to hold shares of up to 50% of the agricultural business.[23] With regard to the leasing of farmland the revision included the following specific changes:

1. Leasing of farmland
   (a) All lease contracts must state that the agreement will be void if the land is not properly used for agricultural purposes.
   (b) At least one member of the corporation must be involved in farming on a full-time basis.
   (c) The corporation must participate in regional agricultural meetings and activities for maintaining infrastructure such as farm roads and waterways.

2. Sale/purchase of farmland
   (a) Individuals may purchase farmland if they are engaging in efficient and effective agricultural activities.
   (b) Purchasing individuals must have a good agricultural management plan that effectively utilizes machinery and labor.
   (c) Purchasing individuals must cooperate with neighboring farmers.

Also included in the amendment of the Agricultural Land Act in 2009, were provisions for investigation, instruction and assignment of right of utilization to reuse abandoned cultivated land.[24]

REVISION OF THE ALL AND FARMLAND BANKS

Landownership laws were then revised again in 2013, when a new institution was created to assist Agricultural Committees in dealing with idle farmland. The reform created the Organization for Temporary Farmland Management (OTFM) to which idle land would be leased and then either rented to farmers or corporations looking for land. This process streamlines the way idle land is dealt with, as under the previous policy, the often overburdened agricultural committees would consult first with owners when farmland was either idle or was underused and then categorized as "renewable" or nonrenewable," if the land was declared "nonrenewable" then the AC would petition the prefectural governor to remove it from the registered farmland book. The reform streamlines this process and allows the OTFM to make decisions about when land is transferred and how. In addition to the OTFM, the MAFF also created the Farmland Intermediary Administration Organization (FIAO), an intermediate or "middle-man" program, to promote leasing of farmland to help encourage more farm consolidation. Coinciding with these changes to agricultural land laws, farmers and nonfarming entities are establishing more agricultural corporations which can help improve efficiency and profitability.

A total of 1071 companies have launched food businesses since the ALL was revised in 2009, allowing corporations to rent farmland across the country. The FIAO also aims to increase farm size through several steps, including renting land from farmers/landowners and redistricting small farms to form larger plots in order to lend it to large-scale commercial farmers, such as agricultural corporations and business enterprises. According to MAFF's most recent data (2010), full-time commercial farmers operate about 50% of Japan's total arable land, and the GOJ aims to increase this to 80% within the next ten years. Japan also expects the FIAO to restore abandoned farmland (400,000 hectares as of 2010) and prevent more of Japan's countryside from being abandoned.[25]

This most recent set of reforms also addresses the Agricultural Committee, the local/municipal institution responsible for implementing the ALL. The reform of the Agricultural Committee, regulates changes to farmland use policy under the Agricultural Land Act. Policies concerning the lease of farmlands are outlined together with revision of the ALL which suggests the strength of the relationship of the two matters. Agricultural Committees are the local government actors responsible for policy implementation, and until recently they were mostly comprised of local farmers. According to recent policy, however, agricultural committee members instead of being voted on by local farmers are appointed according to the selection of local mayors. Moreover, the committee members' numbers have been reduced by half and their role in farmland management has been reduced. As described in earlier sections of the chapter, the agricultural committees oversaw land transfers in the past, but according to the recent reforms under P.M. Abe this job will now be done by the Farmland Consolidation Banks, an institution created by the Farmland Intermediate Administration Organization Law of 2013. The law greatly reduces the power of local agricultural committees by authorizing the transfer of farmland to Farmland Banks which are organized at the prefectural level.

The revision of the ALL in 2013 and in 2014 and adoption of the FIMO law outlines countermeasures against abandoned cultivated farmlands and regulates farmland conversion through the reform of agricultural committees, establishment of Farmland Banks and changes to laws regarding corporate leasing and ownership of farmland.

In 2013, Farmland Consolidation Banks or Farmland Banks were established in each prefecture to promote the consolidation of farmland. The banks are intermediate institutions which oversee the transfer of idle or abandoned farmland and manage local farmland transactions, taking over the primary task of agricultural committees. The banks are authorized to borrow uncultivated, fallow land from owners and lease it in bundles to farmers that are willing to expand the scale of their production (so-called business-minded farmers or business farmers, according to the MAFF). The banks have land utilization rights under the ALL reforms, as farmers relinquish their right to control to whom their land is leased. The MAFF contends that the banks are designed to assist with initiatives that respond to the needs of new companies entering into agriculture, to oversee initiatives related to increasing agricultural corporations (which increase production) and promote farmland consolidation. The banks help those private agricultural corporations wishing to take up agriculture because leasing land was difficult for nonagricultural corporations as their activity was restricted by the ALL. The reform of the ALL which created the Farmland Banks also created a new system that makes the allocation of land public. Those intending to lease land, must subscribe to a publicly announced list, public registers of local farmland are required as well. This theoretically means that information on all available farmland for lease across the country is available and accessible by private corporations. The Farmland Banks have the potential to significantly restructure agriculture in Japan depending on a number of key issues which may limit their impact. Foremost among these issues is the degree to which local-level actors have the ability to manipulate the system.

As Jentzsch[26] (2017) notes, local actors can perform what he calls "defensive consolidation." This term refers to efforts by local stakeholders, including local Japan Agriculture or JA, cooperatives and local farm households which can restructure themselves as larger collective farms. In his research, he notes that these local actors have rendered corporate access from the outside virtually impossible by corporatizing local farms as agricultural production companies preventing access by other entrepreneurs. The hamlet collectivized farms are one of the categories of farms that is included in the reforms privileging agricultural production corporations, but these farms merely incorporate local control as they are often run by a group of retirees or part-time farmers. Therefore, "(I)mplementing farmland consolidation and corporatization by promoting collective farming is thus an opportunity for the co-op (and the local government) to secure control over the local farm sector" (Jentzsch 2017, 41).[27]

Consolidation through collective hamlets such as those described by Jentzsch also preserves local access to state subsidies and maintains the current power structure of JA. Other authors have noted the weaknesses of the Farmland Banks, primarily because of the incentive for farmers to retain ownership over their land either for continued access to government subsidies (when part of an Exclusively Agricultural Area, EAA) or for the potential profit of selling their land for nonfarm uses as outlined in detail by Godo. The potential profit that is present for selling land to development is much more attractive to most farmers than giving up their land to the Farmland Banks for consolidation. Moreover, the Income Support Direct Payment Plan or gentan is still substantial even though it has altered its form, providing incentives for farmers to maintain ownership of idle land that qualifies them for the new direct payment system that pays landowners for maintaining irrigation systems and roads to farms. These forms of payments are also allowable under the current regime of free trade, because they are not a direct form of income support, but they would still be included in Japan's Producer Subsidy Equivalent (PSE) measure. While these incentives continue to exist, both the consolidation of farmland and the restoration of dilapidated farmland to active farmland are unlikely to become a widespread phenomenon that alters the structure of landholding in Japan substantially. As noted in a Yomiuri Shimbun survey, the current targets for the consolidation of farmland through Farmland Banks have been far from the government target. In 2014, the government set a goal of consolidation 130,000–140,000 hectares of land through the banks, the actual amount of land leased by the banks was only 12,400 ha.[28]

The redistribution of profits from farmland conversion to regions and municipalities (when farmland is owned by local governments) addresses some of the core weaknesses in existing law regarding farm ownership. There is potential for large-scale redistribution of farmland and for significant profits which may contribute to a healthier agricultural market and lead to increased sales of farmland which would promote conversion as well. An official with the MAFF described the consolidation of smaller paddies into larger ones and the redistribution of farmland as "the most important change" required for Japanese farming to modernize.[29] Currently, the system for transferring farmland is maintained at the prefectural and municipal level. Efforts to redistribute farmland and encourage the shift to large-scale corporate farming by relaxing the regulations on the use and ownership of farmland depends upon local level actors, especially members of the cooperatives of *nokyo* who have a vested interest in maintaining the system of land-ownership as it exists. The persistence of cooperative power at the local, prefectural and national level although targeted by reforms, has yet to diminish. While JA was reformed, losing its ability to audit and advise local JA's the new audit firm that most local JA cooperatives are using is a spin-off from JA Zenchu. The local efforts to defensively consolidate land in advance of the entry of new farmers and privately owned corporations in the case presented by Jentzsch was overseen by local level Japan Agriculture cooperatives whose strength and ability to adapt to government regulations to protect their interests shows no signs of diminishing.

It is still too early to predict the impact of these changes on the rice-growing culture and the domestic groups including rice growers, JA Zenchu and local cooperatives. The recent policy directives of the MAFF illustrate a concern with maintaining the preservation of rice as a key commodity and promoting Japan's traditional *washoku* diet with rice as the central feature to the world is the central focus of these policy directives. Although the MAFF has made a goal of promoting farmland consolidation through its Core Farmer program and the Farmland Banks, the evidence noted above suggests that the rice-growing community as it currently exists is more interested in preserving local control over land as well as the system of growing rice in smaller plots. It remains to be seen whether corporate ownership will begin to penetrate these landowning patterns and consolidate land into larger farms bringing with it large-scale industrial agricultural businesses. Certainly, international trade agreements will affect how these relationships will be affected by future negotiations and the ability of the MAFF to keep support of farmers in place and limits on rice imports in place, this topic is the focus of the following chapter. The tension between the constituencies in Japan's bureaucracy may also play a role, if the norinzoku advocates hold fewer positions of power in the key agencies, they will lose their power to advocate for control over these policy areas and in time, their budgetary power may lessen. These changes are likely to take decades to occur and in terms of the shape of the regions international trade, the slow pace of change in Japan may diminish its ability to influence the region with the strength that it once had.

Notes

1. See Curtis (1989) and Pye (1985).
2. Kawanaka, N. 1974. "Nihon ni okeru seisaku kettai no seiji katei" [The Political Process of Policymaking in Japan]. In *Fendai Gyosei to Kanryosei*, ed. Ken Taniuchi, et al. Tokyo: Tokyo University Press.
3. Curtis, Gerald. 1989. *The Japanese Way of Politics*. New York: Columbia University Press.
4. Pempel, T. J. 2006. "A Decade of Political Torpor: When Political Logic Trumps Economic Rationality." In *Beyond Japan: The Dynamics of East Asian Regionalism*, ed. Peter J. Katzenstein and Takashi Shiraishi, 35–62. Ithaca, NY: Cornell University Press.
5. George-Mulgan, Aurelia. 2005. "Where Tradition Meets Change: Japan's Agricultural Politics in Transition." *The Journal of Japanese Studies*. 31 (Summer, 2): 261–298.
6. Johnson, Chalmers. *MITI and the Japanese Miracle.*
7. George-Mulgan, Aurelia. 2005. *Japan's Interventionist State: The Role of the MAFF*, 8. London: Routledge.
8. Department of Foreign Affairs and Trade. Draft, n.d. "Agriculture and Food." In *Japan at the Crossroads: Strategies for the 21st Century.*
9. Godo, Yoshihisa. 2007. "The Puzzle of Small Farming in Japan." Asia Pacific Economic Papers, No. 365. Australia-Japan Resource Center, ANU College of Asia and the Pacific, Canberra, Australia.
10. Takahashi, Diasuke. 2012. "The Distributional Effect of the Rice Policy in Japan, 1986–2010." *Food Policy* 37: 679–689.
11. Ministry of Agriculture, Fisheries, and Forestry, Government of Japan. 2015 (April). *Summary of the Basic Plan for Food, Agriculture, and Rural*

Areas: Food, Agriculture and Rural Areas Over the Next 10 Years. Tokyo: Ministry of Agriculture, Fisheries and Forestry.

12. George-Mulgan, Aurelia. 2005. *Japan's Interventionist State: The Role of the MAFF.* London: RoutledgeCurzon.

13. George-Mulgan, Aurelia. 2006. *Japan's Agricultural Policy Regime.* New York, NY: Routledge.

14. Using cheap Thai rice (40/kg) as a benchmark compared to high priced California rice (100/kg), Japan would be allowed to tariff California rice at 300%. With domestic rice at 300/kg and California rice tariffed to 400/kg, the tariff would allow Japanese rice to be competitive with California rice. If Japan waited to tariff after 2000, the tariff rate would have to start from 15% lower than 1995, possibly calculated according to a new method. By terrifying early, the MAFF would be able to set the level high enough to keep foreign rice uncompetitive (Mulgan 2006, 106).

15. George-Mulgan, Aurelia. 2005. *Japan's Interventionist State: The Role of the MAFF.* London: RoutledgeCurzon.

16. See Timothy George (2001), Broadbent (1997), LeBlanc (1999), Maruyama (2003), Penny (2011), and Schreurs (2002) for detailed cases and further analysis of this broad issue.

17. Ministry of Agriculture, Fisheries and Forestry. 1949. Basic Law on Food, Agriculture, and Rural Areas. Available at http://www.maff.go.jp/e/policies/law_plan/basiclaw_agri.html. Accessed February 28, 2017.

18. George-Mulgan, Aurelia. 2005. *Japan's Interventionist State: The Role of the MAFF.* London: RoutledgeCurzon.

19. Davis, Christina and Jennifer Oh. 2007. "Repeal of the Rice Laws in Japan: The Role of International Pressure to Overcome Vested Interests." *Comparative Politics* 40 (1): 21–40.

20. Ministry of Justice, Government of Japan. 1952. Agricultural Land Act. Japanese Law Translation Database. Available at http://www.japaneselawtranslation.go.jp/law/detail/?id=2839&vm=04&re=02&new=1. Accessed February 28, 2018.

21. MAFF. 2015 (April). Summary of the Basic Plan for Food, Agriculture, and Rural Areas: Food, Agriculture and Rural Areas Over the Next 10 Years.

22. Godo, Yoshihisa. 2007. "The Puzzle of Small Farming in Japan." Asia Pacific Economic Papers, No. 365, Australia-Japan Resource Center, ANU College of Asia and the Pacific, Canberra, Australia.

23. Clever, Jennifer, Midori Iijima and Benjamin Petlock. 2014. "Agricultural Corporations Help Revitalize Japan's Farm Sector." GAIN Report #JA4019, *Global Agricultural Information Network.* Tokyo, Japan: USDA.

24. Ibid.
25. Hori, Chizu. 2015. "Ongoing Agricultural Reforms Led by Abe Administration." *Mizuho Economic Outlook and Analysis.* Tokyo, Japan: Mizuho Research Institute, Ltd.
26. Jentzsch, Hanno. 2017. "Abandoned Land, Corporate Farming, and Farmlandbanks: A Local Perspective on the Process of Deregulating and Redistributing Farmland in Japan." *Contemporary Japan* 29 (1): 31–46.
27. Ibid., 41.
28. Hori, Chizu. 2015. "Ongoing Agricultural Reforms Led by Abe Administration." *Mizuho Economic Outlook and Analysis.* Tokyo, Japan: Mizuho Research Institute, Ltd.
29. Interview. 2017. By the author, with MAFF official, July 26.

CHAPTER 3

Global Orders of Trade: Pacific Partnerships and International Agreements

The so-called global order of trade as we know it is relatively young, beginning at the end of World War II with the Bretton Woods institutions. These institutions contain a number of key concepts for ordering trade relationships, initially between the United States and Europe and over time expanding as the number of newly industrialized countries, countries emerging from colonization and post-Soviet block nation-states joined its agreements. The foundation for these trade relationships was the General Agreement on Tariffs and Trade or GATT, which became the World Trade Organization in 1995. The former GATT and the new(er) WTO are governed by 3 core principles reflecting the concerns of the victors of WWII. These three principles are stable currencies, most-favored nation treatment for all countries and decreasing trade barriers over time. These core principles do not acknowledge differences in power amongst its members or difficulties with access to the market because of a lack of technology or any other access issue.

Both of these agreements represent a restructuring of the global economy, in a number of areas deemed vital to its proponents. These areas have not changed significantly over time, but the rules governing them have become broader and more intense, with greater detail and specifity. Also, the rules are now linked with a fine-based system of compliance. In particular, the GATT and WTO focus on a rules based system of trade where forms of protectionism are discouraged. One could argue persuasively that the global food system exists because of these agreements. Those who analyze interactions between states and the forces of global

© The Author(s) 2019
N. L. Freiner, *Rice and Agricultural Policies in Japan*,
https://doi.org/10.1007/978-3-319-91430-5_3

capital that make up the food industry argue that states have choices in terms of how they engage with these forces. Corporate actors are beholden to the rules and conditions of states, even as the rules are constantly being reshaped and defined. Therefore, states can transform the globalizing process to their own national objectives. In the Japanese case, this is illustrated very clearly as this chapter will argue. Despite a powerful trade lobby and a highly evolved set of strong international commitments, agricultural protections continue to exist in Japan and to shift.

The GATT and the WTO presume that countries have products that they can trade competitively. The agreements do not acknowledge that competition produces winners and losers or that countries' ability to compete may suffer as the result of being a former colony. The Bretton Woods institutions take relative equality among trade partners as an assumption. When its institutions were set up, the great power system that existed former to WWII was relatively intact (at least in the minds of the institution's founders). The GATT and its supporting institutions (the World Bank, International Monetary Fund or IMF and International Bank of Reconstruction and Development or IBRD) were unlike any other institution in existence at the time of their founding. The WTO (formerly the GATT) in international legal terms is a framework document. Over time frameworks are expected to become more strict, requiring heavier commitments for members and containing stricter penalties for defying them. The WTO has a trade police and a system for resolving disputes that carries legal force through the power of binding resolutions.

The General Agreement on Tariffs and Trade (GATT) did not contain specific provisions regarding agriculture. The main concern of the GATT was to push members to change the bilateral agreements that persisted before WWII in order to even balance of payments problems and create stable currencies resulting from these problems. There is very little mention of agriculture or food in the document. The first mention is indirectly in Article VI on "Anti-Dumping and Countervailing Duties" when taxes are allowed to offset the subsidization or other support of a product coming into a country.[1] The first direct mention is in Article XI, titled "General Elimination of Quantitative Restrictions" which prohibits quotas, import and export licenses that limit the import or export of any product. Part 2a of Article XI exempts temporary restrictions applied to food and Part 2b allows import restrictions on any agriculture or fisheries product so long as they are necessary for the government to protect domestic products.[2] Moreover, under Article XVIII, titled "Government

Assistance to Economic Development", governments are allowed to protect domestic industries when they are related to policies of economic development "designed to raise the general standard of living of the people".[3] Member governments are also allowed under this article to protect domestic industries in order to maintain the establishment of a particular industry. This ability to protect domestic industries for outside products that threaten domestic producers is underscored again in Article XIX (Emergency Action on Imports of Particular Products), which further allows for tariffs along with the quantitative restrictions on imports, which is allowed in Article XI.

Despite the lack of direct mention, governments were allowed in a variety in ways to protect agriculture in the GATT because they are defined as a primary product. In Paragraph 7, the use of a special tax is allowed to protect the price of primary commodities or returning the production of a commodity to domestic producers. Primary products are defined later in Ad Article XVI as "any product of a farm, forest or fishery...".[4] Table 3.1: GATT Measures affecting agriculture below illustrates where, in the GATT, protections on primary products are allowed.

Overall, the focus on the GATT was not on agriculture, but rebuilding economies after WWII and providing money for that rebuilding. The core presumption behind the theory of liberal trade that the GATT rests on is stable currency arrangements, stable currency and balance of trade were the focus of the early meetings.

Over time however, the agreement began to take on an increasing number of issues and during the 1980s especially, less developed countries (LDCs) were vocal about their opinion that the United States, Europe and some Asian countries were allowed forms of protectionism which distorted trade in an unequal way favoring richer countries and leaving the LDC's disadvantaged, unable to even enter the market. These criticisms had the ability to endanger the entire agreement and to

Table 3.1 GATT measures affecting agriculture

Article	Section	Concession
Article XVI	Text of general agreement	Subsidies for primary products
Article XXXVI	Trade and development	Improve access for primary products
Article XXXVII	"	Improve access for LDCs
Article XXXVIII	"	Improve access for all primary products

de-legitimize the structure of the GATT itself. The GATT agreement tackled the reduction of trade barriers through the reduction of tariffs, it did not address other forms of protectionism. During the Uruguay Round (from July 1986 to November 1992) the members were forced to tackle protections for agriculture or the credibility of the entire agreement would be lost. Negotiations during that time period focused on a reduction of import barriers, a phased reduction of all direct and indirect subsidies as well as reduction of all measures that impacted import access and export competition.[5] In Asia, the food regime has been an important aspect of trade and the way that Asia interacts with the global market. Japan's annexation of Taiwan and Korea was accomplished for the purpose of establishing agricultural colonies, controlling access to natural resources in Manchuria and South East Asia. The deepening dependency on foodstuffs is part of Japan's interaction with its Asian neighbors, it is dependent on Thailand for rice, food and poultry and on China for fish and vegetables.[6] After WWII both Japan and South Korea were early recipients of US food aid, over time these relationships have become multilateral. Japan undertook land reforms and altered farm policies to support industrialization, their domestic production was complemented by imports from the United States and South East Asia. The way that Japan understands its relationship with the world in terms of global trade is heavily shaped by agricultural forces and its reliance on imported food.

Over the years, these forms of protectionism did allow members using them to enjoy privileged access to agricultural markets but the costs added up over time and there were budgetary pressures to bring these costly subsidies under control in both Europe and the United States. Also, the forms of protectionism often involved the stockpiling of grain (especially corn and rice) and these huge surpluses were a problem as well. The Uruguay Round focused on using a measure from the Organization for Economic Cooperation and Development (OECD) called the producer subsidy equivalent (PSE). The PSE measures price support, direct payments, input subsidies and indirect subsidies (regional support and tax breaks), then this number is compared to the net production of each country as an average over time. By the mid-1980s the PSE was substantial for the United States, the European Community, Australia, Canada, and Japan. For Japan the number averaged at 72, meaning the Japanese subsidized 72% of net production. The economic tools available to these countries came with high budgetary costs, this level of budgetary spending simply could not be matched by LDCs who could use the less costly measure of import restrictions and tariffs (which would generate income) but these were the exact tools prohibited by the GATT. The negotiations over agriculture issues were lengthy and resulted in what is known as the Dunkel/Washington accord, a compromise created by merging a proposal by then secretary General Dunkel and an agreement between Washington and the European Community (EC) known as the Washington Accord. The Dunkel Agreement asked members to decrease tariffs, including non-tariff barriers by 36% with a minimum reduction of 15% in each tariff line.[7] Also, members would reduce their aggregate measure of support (AMS) by 20% and decrease budgetary outlays on export subsidies by 36%, reducing the total subsidized export volume to 24%. The Washington Accord amended by Dunkel Agreement by including a 10% EC preference margin instead of 36%, the EC agreed to conform to the domestic support numbers except in crops and livestock head-age. The budgetary restrictions would remain, but the EC would be allowed to continue payments under the Common Agricultural Policy (CAP) even though they met the condition for restriction under the Dunkel Agreement.[8]

This compromise was politically sensitive, but it allowed the EC to retain some control and to preserve the CAP, which was also under negotiation during the Uruguay Round and the EC was involved in those sticky negotiations at the same time. In terms of the PSE, the EC was a minimal offender compared to Japan. The compromise reached between the United States and EC (the Washington/Dunkel agreement) became the basis for the World Trade Organization Agreement on Agriculture or AOA.[9] The AOA was folded into the last round of GATT negotiations (the Uruguay Round) in 1994 when the GATT was in the process of becoming the WTO. Although some make a distinction between the Uruguay Round Agreement on Agriculture (URAA) and the AOA, they are in essence equivalent.[10] The URAA was designed to tighten the regulations on the support the member countries were allowed to give to agricultural producers and the compromise reached during this round became the AOA as explained in Table 3.2: *Dunkel/Washington agreement and AOA*.

In each trade focus category, the AOA is based on the Dunkel Accord, however it includes some special exceptions that limit the AOA's commitments for developed countries. The AOA is divided into several parts, the first is the actual text of the agreement, the second lists the country schedules and the third lists the technical details or modalities that

Table 3.2 Dunkel/Washington agreement and AOA[11, 12]

Trade focus	Dunkel/Washington	AOA	Annex 2 members
Domestic support	Reduce AMS 20%	Reduce AMS 20%[a]	Reduce by 13.3%
Market access	Decrease NTBs 36%	Decrease NTBs 36%, 3% of consumption 5% by 2000	Decrease by 24%, 1% rising to 4% by 2004
Export subsidy	Reduce export subsidies 36% reduce total subsidized exports 24%	Reduce export subsidies 36% reduction of total export volume 21%	Reduce by 24% reduction of total export volume by 14%

[a]Reductions do not include so called "green box" measures those not subject to trade commitments that are said to be minimally trade distorting, such as research, extension, food security stocks, disaster payments, and structural adjustment programmes

pertain to each individual commodity and the country commitments. Here, I will present details concerning the actual text of the agreement before discussing Japan's country schedule as it relates to rice. As presented above, the three areas of focus for the AOA are market access, domestic subsidies and export subsidies. For Japanese rice, the sections on market access and domestic support have the most impact. In terms of market access, the agreement limits tariffs and non-tariff barriers. All existing tariffs are bound and all non-tariff barriers must be converted to tariffs, this is called tariffication. The tariffication should be equal to the barriers in place during the base period from 1986 to 1988. Furthermore, the tariffication is a 36% un-weighted average reduction with two exceptions. The first is the Special Safeguards Provision, the second is the Special Treatment Clause, otherwise known as the Rice Clause.

The Rice Clause allowed for a post-phonement of a country's tariffication until the end of the tariffication period in 2000. Developed countries (like Japan) agreed to grant minimum access of 4% of the amount during the base period of domestic consumption during the first year of implementation and this would rise to 8% by 2000.[13] In order for countries to be granted the Special Treatment Clause they had to declare it in their schedules (4 countries made use of the clause: Japan, South Korea, the Philippines, and Israel). Three conditions of the clause had to be met in order for a country to qualify. First, the import of the good(s) in the base period had to be less than 3% of domestic consumption. Second, export subsidies could not be provided during the base period. Third, the good had to be a predominant staple in the traditional diet of the country. Also, if a country was to receive special treatment under the clause after the agreement was implemented it had to give "additional and acceptable" concessions which would be determined by negotiation. In this case the tariff applied would be equal to the 1986–1988 base period, less 15%.

The domestic support measures of the AOA have been categorized by researchers of the AOA into three categories or boxes relating to how the agreement allows the support. The green box contains support that is considered acceptable because it does not distort trade significantly. The blue box lists levels of support that are set aside as special and do not require reduction. The third or amber box list the most trade distorting domestic support measures that must be reduced. Domestic support in the WTO AOA is measured by the Aggregate Measure of Support (AMS), the AMS only includes direct payments made by governments to producers and does not include any indirect support. In the area of domestic support, agriculture is given special protection according to the de minimus clause 5% of a country's total agricultural production.

The special programs set off in the blue box were argued for by the United States and European Community who created production limiting programs to maintain payments to farmers. These payments take several forms: (1) Payments based on fixed areas and yields; (2) payments made on 85% or below of base levels of production; and (3) livestock payments based on a fixed number of head. These payments must abide by the due restraint clause, meaning they can't exceed the level of support given in 1992. In 1996, the United States altered the way that it gives domestic support to farmers through the US Federal Agriculture Improvement and Reform Act (FAIR) replacing direct payments to farmers with a production flexibility contract program. These contracts are a loophole to the blue box and amber box limits to domestic support because they meet the requirements of the green box. These payments are not paid for by consumers, they come out of the budget approved by Congress and they are not a price support to producers (which would act as a mechanism to lower prices to consumers and heavily distort trade, according to WTO rules). However, these measures do distort trade because they give supported producers and the products they produce an advantage. Farmers are receiving payments that support their income and alleviate the cost of farming a certain good, thereby making that good less costly that it would be otherwise. These payments also prop up those industries that produce those goods, insulating them from full competition.

Chart 3.1 lists those policies that the WTO defines as *minimally* trade distorting and sets off as distinct from those policies that are *heavily* trade distorting. The reason that these indirect payments are not considered trade distorting stems from the way that trade distortion is measured in the WTO, which relies on the AMS (aggregate measure of support) instead of using the producer subsidy equivalent used by the OECD discussed earlier. The PSE is in fact a much broader measure of trade distorting policies as it takes into account both direct and indirect forms of support and all monies from government programs for producers. The PSE includes many of the policies listed in the box above, which do have an impact on trade, whether it is recognized by the WTO AOA or not. The United States, Japan, and the European Union (EU) have continued to support agriculture with a variety of programs, and that support has not diminished, even given their commitments to the AOA which had a stated goal of limiting all forms of protectionism to implement fair and open markets.

As shown in the Table 3.3, levels of support for agriculture have continued to be high for the largest grain producers in the world and Japan. The Japanese PSE is the highest of all countries and has not deviated significantly despite commitments made to the WTO AOA.

Chart 3.1 List of allowable domestic support measures for producers

1. Government service programs
   a. Research
   b. Pest and disease control
   c. Training
   d. Extension and advisory programs
   e. Inspection
   f. Marketing and promotion
   g. Infrastructure
   h. Infrastructure associated with environmental protection
2. Public stockholding for food security
3. Direct payments to producers
4. Decoupled income support
5. Insurance programs and safety net programs

Structural Adjustment assistance

1. Producer retirement
2. Resource retirement
3. Investment Aids
4. Environmental programs
5. Regional assistance programs

Table 3.3 Japan's Producer Subsidy Equivalent (PSE)[14]

Country	2000	2001	2002	2003	2004	2005	2006	2007	2008	2009	2010	2011	2012	2013	2014	2015
Australia	3.74	3.5	4.91	3.66	3.39	3.64	4.41	4.83	4.42	3.09	2.95	3.12	1.96	2.06	1.45	1.34
Canada	19.38	15.61	20.3	24.19	20.26	21.18	20.9	16.43	13.51	17.2	16.44	14.88	14.12	10.13	9.64	9.4
Japan	59.73	56.32	17.18	57.42	55.92	53.79	51.5	46.58	48.31	48.61	54.55	51.3	55.04	52.17	49.5	43.07
USA	22.66	21.43	17.85	14.74	16.04	15.04	11.1	9.67	8.62	10.1	8.57	8.01	8.45	6.9	10	9.44
European Union	32.87	30.28	33.9	33.78	32.98	30.76	29.13	23.24	22.95	23.35	20.04	18.19	19.38	20.06	18.11	18.92

Despite the commitments made by countries to open and liberal trade, it is clear from examining the WTO AOA and the policies of its most powerful members, that this pledge is not meaningful. As countries have acceded to the AOA, they have also become more adept at creating policies that skirt the rules so that they mesh with the allowable measures of domestic support listed in the green box in Chart 3.1 above. For example, the United States has increased its own insurance programs for farmers and decoupled (separated) income support payments to farmers (from yearly production). Researchers who have critiqued these sorts of measures note that although the programs are not linked to production, they do in fact have an impact on production costs and represent a domestic form of support that is very costly and difficult for developing countries to match.

Since the AOA went into force, Japan (like the United States) has also increased its efforts to support domestic agriculture in the form of green box programs that are exempt from their global trade commitments. In the most recent Annual Report on Food, Agriculture, and Rural Areas, the MAFF outlines a number of programs and many of these are green box exempt supports. One of the biggest set of initiatives being promoted by the MAFF currently are those aimed at restructuring agriculture through the promotion of larger paddy fields by "accelerating farmland concentration and intensification". Land improvement acts as a form of decoupled income support for rice growers as the government subsidizes farmers to grow other crops instead of rice. These payments are not linked with production of a specific good and a subsequent payment because of loss of income because the price of that good has fallen. Rather, the government is paying farmers to grow crops that may be more profitable than rice. Some of these payments can also be made under allowable structural adjustment assistance green box measures as resource retirement. The MAFF is also directing energy to programs that attract and train younger farmers to take up farming through farming grants for young farmers and an agricultural employment program, which creates a "full-time counseling service for those who wish to engage in farming, and the provision of farming workshops in agricultural corporations".[15] This type of program providing training, extension and advisory assistance and marketing and promotion is also included in the list of allowable government service programs in the green box in Chart 3.1 above. Another initiative is the expansion of insurance and safety net programs for retiring farmers and

(A)n income insurance system, which can provide a comprehensive service to the entire income of an agricultural business owner regardless of product category, will be established as a safety net for agricultural business owners who work on their management development based on a free management decision.[16]

Essentially the revenue insurance system is another form of decoupled income support that compensates farmers against agriculture losses, which makes farming profitable even when there is a decline in price on the market thereby insulating farmers from the market itself.

All of the programs outlined in the most recent MAFF policy outline are green box programs that are exempt from Japan's commitments to the AOA as well as more recent commitments to agreements like the CPTPP. They can either be construed as government service programs, forms of decoupled income support, insurance and safety net programs or they are a form of structural adjustment assistance. Please see Chart 3.1 for a complete list of policies and the coinciding green box form of allowable assistance with which the policy conforms.

US-JAPAN RELATIONSHIP

Despite the fact that the WTO is a multilateral arrangement, bilateral agreements among trade partners still exist. In Asia, the most important of these relationships is the one between Japan and the United States that is seen by both countries as one aspect of a complicated multi-faceted relationship including security and military aspects and diplomacy. Current trade policy between the United States and Japan then must be viewed within the context of all of the other aspects of their relationships. At times, this relationship has predicted US trade behavior toward its other allies and at times it presents as a contrast to US trade policy toward international actors and other trade partners. After Trump's election in 2017 and his announcement that the United States would withdraw from the Trans-Pacific Partnership negotiations, Prime Minister Abe made moves to enter into other FTAs with the European Union and also revived the TPP negotiations with the other countries included in the agreement before the US withdrawal. Japan's position has been referred to by some as "soft balancing" and an attempt to establish a leadership role for itself in the region expressing itself in terms of diplomatic and soft power.[17] After the United States announced its intention to no longer proceed with the TPP, Japan began negotiating an FTA with the European Union. On December 2017, negotiations were finalized regarding the EU–Japan Economic Partnership Agreement the largest trade agreement in the world. The deal, viewed by some as a rejection of President Trump's more protectionist stance removes some trade duties for Japanese autos entering the European market and lowers trade barriers for European agricultural products (including cheese, wine, beef, and pork).The agreement also reaffirms the commitments of both Japan and the 28 member country EU to the Paris climate accords. The trade deal is roughly the same size as NAFTA in terms of the amount of trade that it covers (about one quarter of global transactions) and following announcement of the deal, the EU and Japan also announced that a strategic partnership on security matters was also in the works.

Following the announcement of the EU–Japan EPA, Japanese trade negotiators also restarted negotiations with the other countries that were part of the TPP, seeking to revive negotiations with those countries under a new trade pact called the Comprehensive Partnership Agreement for the Trans-Pacific Partnership. Unfortunately, despite Japan's desire to move forward, the agreement stumbled in November of 2017 after it the agreement began to receive media attention and Japanese negotiators prematurely attempted to emphasize the importance of reaching a deal. Canadian negotiators however, were unhappy with some elements of the talks casting a shadow over a scheduled ministerial meeting to endorse and announce the main components of the pact among Heads of Government of the 11 remaining member countries. Following its unsuccessful bid to revive the TPP, Japan announced its intention to work on a trade pact with the Baltic countries of Estonia, Latvia, and Lithuania in January 2018, and soon after with Bulgaria as well in advance of the ratification of the EU–Japan economic partnership. Japan's recent strengthening of ties with Europe can be viewed as a balance to China's efforts to firm its ties with Europe through its Belt and Road Initiative, which the country sees as a reviving of the ancient Silk Road trade route with Europe through a number of ambitious transportation projects including a massive rail project connecting China to Europe (a freight line has already been built and is up and running from Xian to Finland). Japan also must consider its strategic interests and the looming threat from North Korea, which it has played up in recent economic talks.

Trans-Pacific Partnerships, Free Trade Agreements and Trade Policy

The Trans-Pacific Partnership (or TPP) included 12 nations (Australia, Brunei, Chile, Malaysia, Mexico, New Zealand, Peru, Singapore, Vietnam, Japan) and the United States before it withdrew, the US was the primary motivator of the trade pact under former President Obama. The agreement had the goal of eliminating 11,000 current tariffs that exist among these countries and to serve as a template for future trade agreements in the region. Getting an agreement rested largely upon the decisions made between the United States and Japan, by far the biggest economies in the deal whose shared trade is seen was a building block of the partnership. The Obama administration viewed the TPP as a key element in its "pivot" or "rebalancing" toward Asia as it sought to counter China.[18] And by Asia, in terms of partners, it really meant Japan, because of the strong US Japan alliance (or *nichibei*, in Japanese) dating from the end of WWII. A report prepared by the Congressional Research Service[19] on key TPP negotiation issues noted the declining US presence in Asia, while America remains "distracted" by the wars in Iraq and Afghanistan. Thus, the TPP was about the US–China rivalry. The concerns of some of the key constituencies affected, including rice farmers and other citizens seemed unlikely to derail the forward progress of the deal, until President Trump assumed office, upsetting the drawn out negotiations, sending the deal into a tailspin.

Japan has played a meaningful role in regional trade as a supporter of the United Nations Food Aid Organization (UNFAO) through aid to developing countries in Southeast Asia as well as spearheading the regionalization process. Japan contributed to the creation of APEC and has encouraged ASEAN in providing a cohesive regional presence. In the postwar era however, Japan's ability to lead a regional trade agreement while limited by regional politics, is also limited by its domestic politics. The economic system propped up by the LDP balances an urban business friendly coalition and a rural, small-business agricultural coalition. This structure has become unbalanced over time as the economy has continued to shrink and big business can no longer afford the cost of maintaining the rural, agricultural structure as it exists and is protected by the MAFF. With this system in place, Japan is unable to respond to demands made by its neighbors and even the United States to reduce barriers to trade for these protected areas. Munakata[20] argues that Japan's pursuit of free trade agreements such as the one with the European Union to improve its oversees prospects and also to promote domestic economic reform to restructure and dismantle this system that keeps Japan from the potential gains that may come with deeper integration, including revitalizing its own economy, capitalizing on Southeast Asia's growth and taking a lead role in the growth of Southeast Asia's economy. As a counterpoint to WTO and global trade, Japan's trade policy includes the negotiation of Free Trade Agreements (FTAs) and Economic Partnership Agreements (EPAs) within the region and beyond. Mulgan[21] argues that Japan's use of these trade agreements makes more sense given their ability to achieve benefits for Japanese business that may also be helpful in the domestic politics that pit business interests against agricultural protection. Essentially, FTAs can be used by Japan's politicians to argue against agricultural protections thereby decreasing the politician strength of the agricultural policy subgovernment. FTA's have distinct advantages for Japan, given what authors like Pekkanen[22] have dubbed "narrow benefits" that also provide more immediate outcomes, especially to business in specific fields such as resource traders, transportation and electronics firms who move commodities that benefit more directly from an FTA than they would from the more diffuse benefits of the WTO. Munakata[23] adds that Japan's pursuit of FTA's reflects a broader political and security agenda. A focal point of this agenda is concern over competition with China over much needed resources (especially energy resources), with China's rise as a regional trade leader and regional integration in the arena of trade that is shaped by Japan. In Asia, the CPATPP, RCEP (Regional Comprehensive Economic Partnership), and FTAAP (Free Trade Area of the Asia Pacific) are large FTAs, also called mega-FTAs that seek to define a rulebook for trade, investment, transportation and other norms that reflect democratic interests while China is pursuing 19 different FTAs including the Belt and Road Initiative (BRI). In 2018, China successfully negotiated FTAs with Notway, Singapore, Pakistan, South Korea and Japan, Panama and ASEAN. Japan has successfully concluded FTA negotiations with 16 countries and is undergoing negotiations with 6 additional countries.

For Japan, the TPP is also part of Prime Minister Shinzo Abe's aim to strengthen the economy under "Abenomics" by boosting the country's GDP through increasing exports to other Asian countries, the US, Canada, and Peru. A boost is certainly needed, as Japan's economy is moving toward a real crisis point as its population ages and economic growth stagnates.

In order to boost the economy and as an attempt to address the most pressing social issues in Japan (the aging population and low birth rate), Abe has been working on his "three arrows" of reform. The three arrows: monetary expansion, fiscal spending flexibility and a growth strategy have been a key part of Abe's platform since 2012. Along with economic reforms, Abe has been juggling the goal of constitutional revisions including changing Article 9 of Japan's constitution, the so-called peace clause which forever renounces military aggression. Many argue that Abe views changing the peace constitution as his ultimate legacy, along with achieving better economic status for his country. The success of Abe's Liberal Democratic Party in the Autumn 2017 snap election in the House of Representatives was seen by some as a mandate for Abe and his continued push for economic reform. Abe's party secured a supermajority, beyond the 2/3 needed required for constitutional changes. Currently, Abe's economic growth strategy has worked to some degree, he has cut the corporate tax rate and reformed corporate governance as well as pursuing trade policies (including the Comprehensive Partnership Agreement for a Trans-Pacific Partnership or CPTPP) that will bring Japan's products a wider audience. However, some criticize the success because it is built on monetary loosening (the Bank of Japan's purchase of government bonds) which is a short-term strategy that ignores the debt with which future generations of Japanese must contend. A more meaningful social security reform will also be necessary soon, in 2025 most of Japan's baby boomers will be 75, they will comprise more than one-third of Japan's population and health care costs will act as a brake on whatever growth the country experiences.

The CPTPP and Other Free Trade Agreements (FTAs)

In some respects, the CPTPP is a work around the recent failure of developed and developing countries to achieve meaningful results at the WTO Doha round of negotiations which began in 2001 and collapsed in earnest in 2008 after a failure by India, China, and the United States to agree on a measure allowing countries to protect farmers during an import surge or price drop. The TPP contains many of the basic provisions that move that Uruguay Round Agreement on Agriculture forward, focusing on eliminating many of the support programs developed countries use to support domestic industries thereby eliminating trade distorting subsides and further opening trade to more equitable treatment in the region. The CPTPP contains sections on Export Subsidies, Market Access, Domestic Support and Export Restrictions. Export subsidies are those subsidies used to support domestic industries, they distort trade by making these supported products cheaper on the global market than they would be without such support. The use of these subsidies has declined significantly since 1995, when the URAA came into effect and the CPTPP eliminates the use of these mechanisms. Market Access in the CPTPP refers to the tariff rates that are put on products entering a country's market, oftentimes these are used to protect specific domestic industries and historically they have been allowed when the tariffs protect those industries that are essential for development or national security. The CPTPP requires member countries to eliminate many of these tariff rates by establishing a quota system for members. Domestic support measures, those government programs used to assist domestic producers have been shifting from amber box measures banned in the WTO AOA to blue and green box programs that are allowed and discussed in detail earlier in the Chapter. Export restrictions refer to methods used by governments to ban the export of certain goods, generally by establishing an export limit or through taxes, thereby shorting global supply. These restrictions are viewed as highly problematic, in the area of food security especially when they are used to distort the trade of global grains such as wheat, corn, and rice. During the 2007–2008 food crisis (discussed at greater length in Chapter 6) when first India, then the Philippines and Indonesia used export restrictions to short the global supply of rice, prices rose 40% as a result.

BASIC OUTLINE OF TRANS-PACIFIC PARTNERSHIP AGREEMENT

For the most part, the current agreement maintains some of Japan's ability to protect its domestic industries, especially beef and rice, despite heavy pressure from the US during earlier bilateral talks. The current deal includes lower tariffs on meat (beef and pork) and rice, considered two of Japan's five "sacred" domestic products (wheat, barley, and sugarcane are the others). Currently Japan imports 770,000 metric tons of rice to meet the rules established by the World Trade Organization. About half of these imports come from the United States, the largest supplier, followed by Australia (another CPTPP partner). Currently rice farmers are protected by Japan's limit on rice imports. According to the deal reached by member countries, Japan's imports of Australian rice will be allowed to rise 6000 tons a year in the initial phase and as much as 8400 tons by the deal's 13th year. Japan will also create a non-tariff import quota for 78,400 tons of rice and a low-tariff quota for milk powder and butter equivalent to 70,000 tons of raw milk. For beef, Japan's current tariff rate of 38.5% will drop to 9% by year 16 of the deal, while those on pork will drop from 482 yen per kilogram to 50 yen per kilo as well (Table 3.4).

From the consumer perspective, these changes are significant. Japan depends on imports for 60% of what they eat, and consumers will see lower costs at the grocery store as a result of this deal. Although Japan made these concessions, it maintained the ability to protect its beef and rice markets in the following ways:

- Japan will be allowed to return tariff barriers in the twentieth year of the agreement for both beef and pork if they see their market flooded by imported beef and pork
- Japan will purchase an equivalent amount of domestic rice to the amount imported to support local producers

Despite these allowances, Japan's rice farmers are still worried about the impact of foreign rice. Farmers that I conducted interviews with in January of 2016 and August of 2017 expressed deep reservation over the agreement. In Joge, a small community near Osaka in Hiroshima prefecture, where many young people have left, houses and land are abandoned. Mr. Mizukami, who is a lifelong resident there, works several small plots of land and told me that the "TPP deal is an important story, farmers like me feel the consequences directly, I'm afraid the farming lifestyle is becoming rare and will disappear".[24] A young female soybean farmer in Joge, outside of Hiroshima, reminisced about a time when active farms and their bright shades of green marked the neighborhood. "One by one they stopped farming, the farms are gone," she said,

Table 3.4 TPP impact on Japanese goods

	Rice imports	Beef tarriff rate (%)	Pork tariff
Current	10,000	38.50	48円
Goal	215,000		
Concession	70,000	9	50円

and now the land has transformed to weeds (Freiner 2015). Mr. Saito, a younger farmer attempting to bring a spirit of entrepreneurship back to the area, fears an influx of foreign rice and landowners and says the TPP and the changes it brings will crush a way of life, and young farmers like him will be unable to survive. "Farmers working by hand can't compete," he said, "We are lost".[25]

CPTPP Negotiations Post-Trump

On October 26th, 2017, Japan entered the final stages of working out a TPP agreement by enacting measures to meet up with the basic prerequisites of the agreement. In the fall of 2017, Japan continued to work with other Asian nations to work out an agreement even after the United States backed out after President Trump took office in January. Japan has played an important role in taking leadership over the agreement and moving forward with the commitments of the now 11 countries that make up the committed countries in the CP (Comprehensive Partnership Agreement) for TPP. Although the CPTPP is not as large as the TPP representing 13.5% of global GDP instead of the TPPs nearly 40%,[26] it is a meaningful accomplishment for those countries who are members. The agreement did suspend some of the provisions most dear to the US, in order to preserve the deal. The most significant of these were the chapters related to intellectual property rights and investment protection provisions. This agreement builds on evolving requirements included in the GATT/WTO, NAFTA and other bilateral agreements over the past 40 years, moving forward with a rules based system of procedures that integrates trade across Asia and the Pacific. The agreement will not enter into force until it is ratified by 6 countries, which will likely happen soon. Japan is predicted to be the first country to ratify, likely after elections in the Fall of 2018. Those provisions discussed below reflect the agreement which was agreed upon by the CPTPP countries. Japan is also taking measures domestically to promote policies in advance of its accession to the agreement and its commitments entering into force. In December of 2017, Prime Minister Abe's Cabinet released a document detailing the importance of moving forward with the TPP for Japan.

In the Japanese government pamphlet written to answer questions about the CPTPP, the argument in favor of the CPTPP mentions 6 characteristics of the agreement. These include building rules for trade, creating a value-chain that is growing in the Asia-Pacific region, contributions to small and medium sized enterprises and local industries, the promotion of new investment, promoting a new future that will create trade standards and investment rules with countries that share universal values and a deepening economic interdependent relationship with countries that share values in the Asia-Pacific that adds to the stability of the region. Although the language is indirect, the agreement certainly reflects a concern with the rise of China (a country with whom Japan does not see itself sharing universal values) as well as North Korea. The agreement could be seen as a soft power mechanism to counter both China's economic expansion in the region and moves that it is making toward creating a trade region based on rules that China has advanced through the Regional Comprehensive Economic Partnership which is proceeding with ASEAN nations (and includes many of those countries that are also members of the CPTPP) as well as its Belt and Road (BRI) initiative. President Trump recognized the importance of this initiative back in May of 2017 when the United States and China signed their own bilateral trade deal which, some argue, showed acceptance of China's leadership in the region and its desire to establish trade norms reflecting China's interests.

The new MAFF program designed to increase Japan's agricultural competitiveness directly addresses Japan's commitment by creating several new policies to strengthen the ability of Japanese farmers and producers to succeed in the trade environment that the CPTPP will create. The major aspects of the policy package include an emphasis on sanitary and phytosanitary (SPS) standards, country and region of origin labeling including information on geographic indicators (GI), new support mechanisms for farmers that meet up with requirements in the CPTPP and new institutions that will promote and market Japanese products overseas that are designed to increase Japanese agricultural exports.

NATIONAL TREATMENT AND MARKET ACCESS

By far, the most lengthy and meaningful section of the CPTPP covers the way in which member countries set policies with regard to the entry of products into their country and the export of products to other countries. Chapter 1 of the agreement deals with these issues with regard to the national treatment of exported goods and market access for imported goods. Section of this Chapter 1, deals specifically with agricultural goods, incorporating Article 2 of the WTO Agreement on Agriculture and also working beyond the WTO to address the issue of export subsidies, export credits, export credit guarantees and insurance programs. With regard to export subsidies, according to the CPTPP, member countries are to provide no export subsidies. The issues of the different forms of export credits and insurances programs (like those provided to farmers in Japan as income insurance) are not prohibited. These sorts of supports are turned over to the Committee on Agricultural Trade which the agreement establishes and countries are urged to "work to develop mult-0lateral disciplines to govern" these programs. The formula for establishing what countries are held to in terms of the elimination of tariffs and tariff rate quotas are governed by this chapter of the CPTPP, however they are extremely complicated and many products have different schedules even for a single country. Japan's commitment will be evaluated with regard to the sacred five products (rice, beef and pork, wheat, dairy, and sugar) that are deemed most important to Japanese culture and society.

Japan's Trade Commitments on the Sacred Five

Japan has committed to reducing tariffs on imported milk and cream by 25% as a baseline number as the agreement enters into force and then to reduce tariffs 16.2% by year 3, 10.4% by year 5 and 7.5% by year 13 forward. For many other dairy products, including milk powder and butter, Japan retains its commitment to the WTO but does not go beyond. Japan's commitments with regard to wheat follows a similar pattern, the ability to extend tariff rates the same as those agreed to in the WTO is applied. The same is true for the tariff reductions with regard to pork (pig carcass). Imported beef whether fresh or frozen is set at a base tax rate of 38.5% and then the tariff schedule diminishes tariffs 24% by year 5, 20% by year 10 and finally drops to 9% by year 16 forward. Sugarcane is treated in the same manner as wheat and pork, following a special formula that allows the country to maintain its ability to tax imported the imported good according to the formula specified in the WTO. For sugarcane, in addition, Japan is allowed to continue an import licensing procedure requiring certification. Many of the abilities that Japan has with regard to controlling the import of rice and wheat is also maintained through the Food Law, which allows the government to purchase and sell staple foods, the MAFF also has the ability to require certification of these products. In side negotiations Japan agreed to allow imports of rice and wheat beyond its commitment for these countries, after the

US backed out, Japan maintained these commitments with Canada and Australia. Japan's imports from Australia for rice are set at Year 1 at 6000 and then reach their highest level in Year 13 at 8400 metric tons. These numbers are significantly lower than the goal of 215,000 that the United States was pushing for in earlier TPP negotiations. The agreement preserves Japan's ability to continue to protect both rice and wheat by limiting their import. Wheat imports are also similar capped in Year 9, for the US at 150,000 and from Australia and Canada at 50,000 and 53,000 respectively. Along with limiting imports of wheat, Japan retains its ability to tax imports. Wheat from Australia, and Canada is taxed in Year 1 at 16% and decreases to 9% in Year 9 and each subsequent year (Table 3.5).

Along with the commitments made to import specific quantities of goods, the CPTPP addresses import taxes as well. For rice, Japan maintains an import tax of 341 yen/kg for rice, a 55 yen/kg tax for imported wheat, a tax on refined sugar of 103.1 yen/kg and for imported pork 524 yen/kg. In addition, Japan will be allowed to tax imported beef at a rate of 38.5%. For rice, the tax is maintained throughout the agreement and is not eliminated. The tax on wheat must be reduced by 45% by the 9th year of the agreement. Both beef and pork were given long term tariff reductions and the MAFF is allowed to enact special safeguard measures if there is a surge of imports. The MAFF is also able to store rice and to use government purchase of domestically produced rice equal to the amount imported to maintain control over its supply as a staple food crop. This control over the supply of rice by government purchase preserves the power of the MAFF and its ability to control the price of rice.

These commitments, while meaningful to some extent, still allow Japan the ability to preserve its control over its sacred products through limiting the amount of goods imported and in some cases also the ability to tax those imports substantially. According to Japan's on Ministry

Table 3.5 Japan's import commitments for rice and wheat

	Rice		Wheat		
	USA	Australia	USA	Australia	Canada
Year 1	50,000	6000	114,000	38,000	40,000
Year 5	54,000	6480	138,000	46,000	48,667
Year 10	64,000	7680	150,000[a]	50,000[a]	53,000[a]
Year 13	70,000	8400			

[a]The quota is for Year 9 and remains at the quantity for each subsequent year

of Agriculture, Fisheries and Forestry, Japan's agricultural industry will not be exposed to substantial competition from imports. In a MAFF pamphlet which answers frequently asked questions about the CPTPP, the Ministry states that only 20% of all agricultural, fishery, and forestry products will be affected by the agreement and the sacred five products will not be threatened by the terms of the agreement at all.[27] The webpage notes that the agreement will demand changes in only 1% of those products considered "sacred".

MODERN BIOTECHNOLOGY

Genetically modified organisms are also dealt with in Chapter 1 of the CPTPP Agreement, under the section on Agricultural Goods. Modern biotechnology is defined in the agreement as the application of science to:

1. In vitro nucleic acid technologies including rDNA direct injection of nucleic acid into cells or organelles or:
2. Fusion of cells beyond the taxonomic family

That overcomes the natural physiological reproductive or recombinant barriers and are not technologies used in natural breeding or selection. This definition follows the standard set by the Cartagena Protocol at the Rio Summit on Biological Diversity in 1993. The Protocol also goes on to define the way in which genetically modified organisms (GMOs) or living modified organisms (LMOs) as they are sometimes referred to are handled in order to preserve human health and safety following the Precautionary Principle of the Rio Declaration which maintains that steps to ensure the absence of environmental consequences must be taken prior to the adoption of new genomic technologies through scientific study. Unfortunately the CPTPP does not go on to discuss how the protocol will be implemented, nor does it apply regulations that apply to GMOs that will be used for consumption and feed or those that will be used for laboratory research as the Protocol does. Rather, the CPTPP does not restrict the trade of GMOs and establishes that countries are not required to modify or adopt laws, regulations or policies to control products of biotechnology. The CPTPP does require parties to make those documents that are required for the production of use of biotechnology publicly available along with the risk or safety assessment that rationalizes the requirement of such authorized usage including a list of products that have been authorized. A working group on products of biotechnology is also established by the agreement. These measures underscore that risks with regard to GMOs are borne by the member countries who must link policies that limit the import of GMOs with scientifically backed risk and safety assessments. Currently, Japan's policies regarding GMOs are fairly strong, they are more heavily regulated than they are in the United States but not as strongly as European regulations as will be discussed at greater length in Chapter 5.

PHYTOSANITARY STANDARDS

The CPTPP preserves the adoption of phytosanitary standards that is outlined in the WTO agreement and for many of the provisions go beyond WTO. Phytosanitary standards were first adopted by the United Nations Food and Agriculture Organization in 1993 as the International Plant Protection Convention. Phytosanitary measures are defined as "any legislation, regulation, or official procedure having the purpose of preventing the introduction or spread of quarantined pests or to limit the economic impact of regulated non-quarantined pests." Phytosanitary measures have the purpose of protecting plant health and preventing the introduction of pests to importing countries while minimizing the interference with international trade. In particular, the CPTPP outlines a more elaborate regime regarding the use of SPS standards as a form of trade protectionism than those of earlier trade agreements, including the WTO AOA. The CPTPP focuses on creating a transparent system for all members, which includes the provision that countries place no restrictions on imported products other than those than are necessary to achieve the sanitary plant quarantine objectives of the member country. Members must report the reasons for quarantine when they occur as a result of import inspection as well as publish standards for quarantine and inspection in advance so that other members can anticipate them and comply. The CPTPP establishes a framework for negotiation that results on consultation with scientific experts called "cooperative technical consultation" when member countries disagree. Unlike the Codex Alimantarius, the international legal document that the CPTPP refers to and is based upon, Chapter 6 of the CPTPP, which deals with Sanitary and Phytosanitary Standards has the primary goal of protecting "human, animal or plant life or health in the territories of the Parties while facilitating and expanding trade by utilizing a variety of means to address and seek to resolve sanitary and phytosanitary issues" (Chapter 7 SPS). This language puts measures to protect human, plant and animal life and health in the context of trade rather than promoting consumer health singularly. The Codex Alimantarius has the purpose of consumer health protection countries should avoid "unjustified differences in the level of consumer health protection" and the goal should be consumer health in this context, not trade liberalization. The CPTPP SPS agreement places consumer health within the context of trade, which potentially puts health standards below trade liberalization, making them subordinate to concerns related to trade.

The SPS aspect of the CPTPP attempts to build in equality similar to those of the Codex in its section on Equivalence, which maintains that member countries should adhere to international standards, guidelines and recommendations when adopting its standards which must be linked to a risk analysis based on scientific principles. In this way, the SPS limits countries ability to create SPS standards that are a form of protectionism thereby distorting trade. The CPTPP goes beyond WTO rules by linking SPS measures with a documented risk analysis that is based on scientific principles and requires countries to make these risk analyses available to other member countries.

The section of Science and Risk Assessment states that SPS standards must be based on "scientific principles" although it does clearly define this term. The standards must not be discriminatory or be applied arbitrarily, the risk analysis must be documented and other parties must be allowed the opportunity to comment. The scientific foundation for these measures must be established and part of international guidelines and recommendations, the guidelines are to be "based on documented and objective scientific evidence that is rationally related to the measure". SPS standards of risk management must not be more "trade restrictive than required unless there is another option reasonably available, taking into account technical and economic feasibility, that achieves the appropriate level of sanitary or phytosanitary protection and is significantly less restrictive to trade" (SPS Ch.7).

Clearly the CPTPP elevates the issue of trade distortion above SPS protections and limits countries abilities to impose these standards. A weakness of the section of science and risk analysis is the lack of a definition of "scientific principles" which raises significant questions. Without a clear definition, it is left open to interpretation and possible manipulation by member countries, especially important is who defines what is and is not a scientific principle and whether or not that definition is left to member countries, their scientific communities or international corporations. Many scientific principles are unpublished and held by private industry as either confidential business information or trade secret. It is un-clear what occurs when scientific principles are not in agreement with one another, as is often the case especially with regard to new scientific developments in plant and animal genomic science. A final major weakness of the section of science and risk analysis is the absence of any mention of genetically modified organisms (GMOs) and the ability of countries to incorporate GMOs into their SPS measures. Final sections on SPS measures that move beyond the standard established by the WTO include those on Audits, Import Checks, and Certification. The section on Audits allows member countries the ability to audit one another's inspection systems, control programs and inspection facilities when they are to import a good that has a SPS measure connected with it by the importing country. With regard to Import Checks, the SPS ensures transparency by requiring that "(A) Party shall make available to another Party, on request, information on its import procedures and its basis for determining the nature and frequency of import checks, including the factors it considers to determine the risks associated with importations" (Ch7SPS).

The CPTPP once again, goes further than the WTO SPS standards by requiring importing parties, to perform these import controls under the auspices of a quality assurance program that is consistent with international laboratory standards and requires that importing parties maintain documentation of test samples, methods to analyze these samples, and findings are commensurate with "available science". With regard to the last import control addressed in the section on SPS standards, certification, the agreement requires that when member countries require certification they must do so "only to the extent necessary to protect human, animal or plant life or health" thereby limiting countries' ability to use certification standards as a form of import control that would distort trade. Information on an importing country must be available to the exporting country and the use of certification must once again be based upon international standards, guidelines, and recommendations.

The second chapter of the CPTPP agreement also deals with rules of origin as well as small and medium sized business. These aspects of the agreement have been noted by Japan's MAFF in its literature that details benefits of the agreement. The rules of origin section explains how goods imported from member counties are to be given preferential treatment according to the origin procedures. Exporters, importers, and producers can claim this preferential treatment by filing the required documentation and maintaining records as required. The rules of origin process includes a mechanism for verification, details the circumstances during which goods may be denied the treatment and the steps countries must follow when denying the preference to another member country's product. The rules of origin preference is intended to create niche markets for local and regional products. Japan's MAFF is making an effort to comply with this aspect of the CPTPP by creating its own special trademark for Japanese products called the JAS Mark.

With both China and Japan attempting to advance economic agendas that will shape the region's trade partnerships, the position of the United States with regard to both countries is an important factor. At this point however, the United States has likely lost its opportunity to help build strong rules and norms in the region which reflect the values that it considers important. The 11 signatories to the CPTPP held a formal signing ceremony on March 8th, 2018. These countries would have to agree to changing major provisions of the agreement to admit the United States, and after Trump first withdrew and then wavered about returning, the other signatories may not be willing to take the risk of sending an agreement which will soon be ratified by member countries back to the negotiating table. The CPTPP negotiations represent a victory for Japan's MAFF, which although making some limited concessions, has largely preserved its ability to protect its rice growers from imports. It is predicted that the US will lose out on trade as Japan's beef and pork market will open up to Australian imports and the new Japan EU FTA will allow European wine in at a lower tariff rate, making it more attractive for consumers than US wine, which will be taxed at a higher rate.

REACTION OF FARMERS, NOKYO, AND CONSUMER GROUPS

Farmers in Japan's largest national organization, JA Zenchu, maintain concern about the impact that imported rice from the United States and other TPP countries will have on Japan's most important traditional crop. Japan's domestic rice production is already threatened by decreasing demand; the addition of foreign competition may be an additional pressure that this long-upheld traditional lifestyle will be unable to bear. The efforts of young entrepreneurial farmers are at the beginning stages in rural communities. The proposed foreign competition from the TPP comes at an especially difficult time in the story of Japan's rural rice-growing lifestyle.

In Japan's case the cost of the agreement involves allowing increased imports of farm products and lowering some tariffs, aspects of the agreement that could significantly harm its farmers, especially those who grow its most important traditional crop: rice. The CUJ (Consumers Union of Japan) and PARC, a non-profit organization committed to international social and economic justice, fear that the TPP and its rules for investment and copyright will open the country to GMO beef and rice from the United States (even though the pact doesn't mention them specifically). Japanese who value local control of their food supply are also dismayed by the potential introduction of more genetically modified organisms. Japan currently cultivates no GMO crops, but the TPP will bring corporate ownership of agricultural land and the inability of government to restrict the use of GMOs. Japan imports many genetically modified foods, but currently do not grow any. Consumers groups, like the Consumers Union of Japan (CUJ), are against any increase in GMO use because of the threat to biodiversity.

Japan's rice farmers have long been the backbone of the ruling Liberal Democratic Party as noted in Chapter 2. But lately, as their numbers dwindle along with a declining population and demand for rice, this key cultural constituency may have lost the strength it once had to demand the government's support. There are now around 2 million rice farmers in Japan, down from 4 million in 1990 and as many as 12 million in 1960. Some farm part-time, while for others it's their entire livelihood and passion. In Toyama prefecture, a region naturally suited to rice growing because the rice paddies here get water from the melting mountain snow, most farmers live a country lifestyle. The farmers I spoke with in Toyama and Hiroshima, areas where small villages dot the landscape, hundreds of miles from Tokyo, said they are most worried about the impact of the TPP on their ability to compete with foreign rice and foreign ownership of agricultural land. Rice is grown in small plots (less than an acre). Currently, it is difficult for outside companies to own land because of legal measures. The TPP would allow for limited foreign land ownership as a form of investment in Japan's rice market but as later chapters illustrate, the nokyo are already acting defensively to prevent foreign investment from occuring. Japan's allowances on rice and other agricultural products to accede to the CPTPP are relatively modest when compared to the demands the US has made, it is likely that Japan will ratify the agreement in 2019 after elections, which favor the LDP and its leader current Prime Minister Abe. It is expected however, that the US may pressure Japan for increased rice imports during bilateral negotiations that Trump and Abe may take up in late 2018 or early 2019. As yet, the TPP will affect farmers and nokyo, although it is unclear how great an impact this will have, the nokyo have shown the ability to adapt to changes and maintain their power. The CPTPP is not the end of the nokyo or of Japanese rice farming, but its affects will be felt.

NOTES

1. General Agreement on Tariffs and Trade. 1986. *Text of the Agreement.* Available at https://www.wto.org/english/docs_e/legal_e/gatt47.pdf. Accessed December 1, 2017.
2. Ibid.
3. Ibid.
4. Ibid.
5. Rayner, A. J., K. A. Ingersent, and R. C. Hine. 1993. "Agriculture in the Uruguay Round: An Assessment." *The Economic Journal* 103 (421) November: 1513–1527.
6. Gordon, Peter Jegi. 1990. "Rice Policy of Japan's LDP: Domestic Trends toward Agreement." *Asian Survey* 31 (10) October: 943–948.
7. Rayner, A. J., K. A. Ingersent, and R. C. Hine. 1993. "Agriculture in the Uruguay Round: An Assessment." *The Economic Journal* 103 (421) November: 1513–1527.
8. Ibid.
9. Kwa, Aileen and Walden Bello. 1998. "Guide to the Agreement on Agriculture: Technicalities and Trade Tricks Explained." Bangkok, Thailand: Focus on the Global South. Available at http://www.jstor.org/stable/10.7864/j.ctt1hfr247.14. Accessed November 30, 2017.
10. GATT. 1993. "The Draft Final Act of the Uruguay Round." *The World Economy* 16: 237–260.
11. Rayner, A. J., K. A. Ingerset and R. C. Hine. 1993. "Agriculture in the Uruguay Round: An Assessment." *The Economic Journal* 103 (421) November: 1513–1527.
12. GATT. 1993. "The Draft Final Act of the Uruguay Round." *The World Economy* 16: 237–260.
13. Ibid.
14. UNFAO, FAOSTAT Food and Agriculture Data, http://www.fao.org/faostat/en/#home. Accessed 11/18/2017.

15. Ministry of Agriculture, Fisheries, and Forestry. 2015 (April). *Summary of the Basic Plan for Food, Agriculture, and Rural Areas: Food, Agriculture and Rural Areas Over the Next 10 Years.* Tokyo: Ministry of Agriculture, Fisheries and Forestry.
16. Ibid.
17. Anderson, Kym. 2017. "Ongoing and Emerging Issues in Agricultural Trade Negotiations." In Kym Anderson, Finishing Global Farm Trade Reform: Implications for Developing Countries (84–96). Adelaide, Australia: University of Adelaide Press. 16. Ibid.
18. Reich, Simon. 2015. "Is TPP About Jobs or China?" *TheConversation. com*, May 22, 2015. Available at https://theconversation.com/is-tpp-about-jobs-or-china-42296. Accessed February 28, 2018.
19. Fergusson, Ian F., Mark A. McMinimy, and Brock R. Williams. 2015. The TPP Negotiations and Issues for Congress. *Congressional Research Service*. March 20, CRS.gov. Available at https://fas.org/sgp/crs/row/R42694.pdf. Accessed February 28, 2018.
20. Munakata, Naoko. 2006. In Peter J. Katzenstein and Takashi Shiraishi eds. *Beyond Japan: The Dynamics of East Asian Regionalism*, 130–160. Ithaca, NY: Cornell University Press.
21. George-Mulgan, Aurelia. 2008. "Japan's FTA Politics and the Problem of Agricultural Trade Liberalisation." *Australian Journal of International Affairs* 62 (2): 164–178.
22. Pekkanen, Saadia M., 2005. "Bilateralism, Multilateralism, or Regionalism? Japan's Trade Forum Choices." *Journal of East Asian Studies* 5(1): 77–103.
23. Munakata, Naoko. 2006. In Peter J. Katzenstein and Takashi Shiraishi eds. *Beyond Japan: The Dynamics of East Asian Regionalism*, 130–160. Ithaca, NY: Cornell University Press.
24. Freiner, Nicole. 2015. "Japan's Sacred Farmers Brace for Pacific Trade Deals Death Sentence." *TheConversation.com*. Available at https://the-conversation.com/japans-sacred-rice-farmers-brace-for-pacific-trade-deals-death-sentence-45280. Accessed October 17, 2017.
25. Ibid.
26. Sherwood, Dave and Felipe Iturrieta. 2018. "Asia-Pacific Nations Sign Sweeping Deal Without the US." *Reuters News*, March 8th. Available at: https://www.reuters.com/article/us-trade-tpp/asia-pacific-nations-sign-sweeping-trade-deal-without-u-s-idUSKCN1GK0JM, Accessed June 19, 2018.
27. MAFF (農林水産省). 2017 (平成29、年10月). "米をめぐる状況について." Pamphlet published by MAFF, Tokyo, Japan.

CHAPTER 4

Japan Agriculture (JA): The Role of the Agricultural Cooperative

It is not an exaggeration to state that JA Zenchu is the largest and most important agricultural cooperative in the world. The JA remains a bulwark of protection (even amidst a forced restructuring because of government policy changes) supposedly advocating for the interests of both large-and small-scale rice growers through a dizzying array of activities that are coordinated nationally at its large office building in downtown Tokyo, one subway stop from government buildings. There is no other organization comparable to JA Zenchu; it coordinates and lobbies for agricultural interests in the realm of politics, it also has its own bank, it controls the infrastructure for storing rice across the country in its large rice houses (sometimes known as country elevators); it even sponsors some of the more popular TV shows. Simply put JA is inescapable. On a train ride from Hiroshima to Toyama, I counted and photographed innumerable local storehouses that were part of its vast network. JA is organized hierarchically, with local cooperatives organized under a prefectural central union called JA Zen-Noh, the prefecture central union is further organized under the national central union, JA Zenchu. The entire cooperative system is organized in three tiers.

At the prefectural level, is JA Zen-Noh, tasked with creating policies to respond both to national directives and to more immediate farmer's concerns and to assist rice growers and manage rice storage and supply. Through its local activities across the country, the organization has

© The Author(s) 2019
N. L. Freiner, *Rice and Agricultural Policies in Japan*,
https://doi.org/10.1007/978-3-319-91430-5_4

unprecedented and singular access to information about rice production, no other national organization has the depth and breadth of activity that can match JA. As the organization states on its own English webpage,

> JA (Japan Agricultural Cooperative) is organized in every prefecture and municipality throughout the country, based on the principle of mutual cooperation, with the purpose of protecting farming and living of its individual members. To this end, JAs are engaged in various activities including farm guidance, marketing of farm products, supplies of production inputs, credit and mutual insurance businesses, while they are referred to as multipurpose agricultural cooperatives.[1]

Although there is no legal provision requiring that farmers be affiliated with Nokyo, virtually all farmers are members and use services provided by its local branches. Unlike agricultural cooperatives in other countries the scope of cooperative activities in Japan is far wider and includes banking, real estate, travel agencies, supermarkets, and even funeral homes. Until recent reforms, the organizations were exempt from antitrust legislation and, as a result, enjoyed monopolistic control over a number of key activities with regard to rice farming (rice storage, transportation and distribution as well as the sale of fertilizers, pesticides, and farm machinery). Because most farmers borrow money from JA at below-market interest rates whenever they buy fertilizers and farm machinery, local branch offices of JA effectively control the fate of each farm household. However, most local organizations managed by JA Zen-Noh have been suffering from chronic deficits relying on Japan's agricultural policy infrastructure to maintain themselves.

HISTORICAL BACKGROUND
OF THE AGRICULTURAL COOPERATIVE

Sangyou kumiai and nokai are the prewar predecessors of the nokyo such as JA. Sangyou kumiai had huge membership (some estimate it at 4 million) during the prewar years and in 1931 were granted a monopoly on the purchase and sale of rice to the government and commercial retailers. This monopoly solidified the organization's power and during wartime, the government combined the sangyou kumiai organizations with the Teikoku Nokai (Imperial Agricultural Association), a national top-down organization that provided education and extension services, to which membership was mandatory in 1905. The newly combined organization was called Nogyokai, this unified organization was under direct control of the state and was used to help finance the efforts and to begin the process of land reform by circumventing the old feudal payment scheme that went from tenant farmers to landlords. The Nogyokai organized direct payments to farmers for their produce, cutting the landlords out of the process. The Nogyokai essentially became the "new" nokyo after the war. In terms of postwar reforms, SCAP (the Supreme Commander for Allied Powers, Douglas MacArthur) was most concerned with achieving land reform and was not concerned with the sort of organization that would replace the old Nogyokai. What emerged then was more a preference of Japanese politicians, who desired minimal reforms and basically just renamed the Nogyokai system (what critics called *kanban nurikai* "switching the signboards").[2] Likely, the main reason the organizations became so powerful was that the prewar Nogyokai was the only organization with the resources and structure that had the ability to assist with the postwar food shortages by gathering food and distributing it throughout the country. At the end of WWII, when the government first began to end the system of price controls for staple goods, Japan's agricultural cooperatives made one of their first demands on the new government. These demands would predict future directives of cooperatives as Ronald Dore notes this early encounter in his landmark study *Land Reform in Japan,*

> In September, 200 representatives of the Agricultural Cooperatives and Prefectural Agricultural Committees held a meeting which issued a strongly worded condemnation of reckless suggestions in government quarters that the current (price control) system be changed. These suggestions, the statement said, were a defilement of the farmer's efforts to operate the new voluntary pre-contract system of rice deliveries, and showed a flagrant disregard of the wishes of the consumer.[3]

Dore suggests that the fate of the Cooperative agencies is bound up with the continuance of price controls and that the system of assistance for agriculture after the war had a single primary aim: increasing Japan's food self-sufficiency. The rationale for this aim is understandable given the context of Japan's food supply prior to and during the War when shortages racked the economy and consumers sometimes became rioters, destabilizing the regime. As explained in Introduction, the agricultural cooperatives that emerged after the war were linked to efforts by the left wing, and the farmers and tenancy movements in particular. During the war effort, these organizations were brought under greater government control and folded into government supported agricultural associations. The associations existed in every village and membership was compulsory, as the government needed channels to monitor the food supply and to ration it amongst citizens. Therefore, the Cooperatives that were reformed after the Agricultural Cooperatives Law of 1947 have an ingrained history of being affiliated with the government as part of its staple food control system.

The preoccupation with food self-sufficiency may have had an understandable rationale then, but it is surely less relevant now as storehouses of rice illustrate over the abundant availability of Japan's staple food. After WWII, cooperatives were relatively weak organizations with little working capital, dependent on funding from the government. Moreover, Dore[4] notes that there was little enthusiasm or loyalty for the organizations outside of loyalty to the village. The involvement in the rice supply and control of the rice supply and its supporting apparatus was the most powerful feature of cooperatives after the war and this continues today. The cooperatives profited from handling and storage charges, the coverage of losses through crop insurance and through Bank of Japan loans for agricultural bills given to farmers to secure the rice crop.[5] Cooperatives also provided credit to farmers and could deduct loan repayments from farmer's incomes; making loans to farmers allowed the cooperatives the ability to handle fertilizer sales.

These activities remain part of the core source of power for JA even today. While, it has agencies that provide health and welfare as well as press and promotion activities, the lion's share of financing for farmers is from its bank, insurance organization, and supply and distribution of rice through JA Zen-Noh (Prefectural Central Union of Agricultural Cooperatives).

Garon[6] includes activities of the early agricultural cooperatives in his discussion of the mobilization of national power in post-WWII Japan and the array of institutions focused on popular welfare. The predecessors of today's JA, the *sangyou kumiai* were state regulated and functioned as a source of credit for farmers and then became affiliated with the New Life campaigns of the 1950s which encouraged savings and frugality. Agricultural cooperatives then have links to these moral suasion campaigns which urged Japanese to increase productivity and increase their savings which is thought to assist with price stability and allow the population to remain less dependent on government social programs. The promotion of savings and the use of credit agencies such as the agricultural cooperatives are part of the state's efforts to use its agencies to influence the behaviour of its people with the goal of strengthening savings and as Garon states these campaigns have "brought together government and private organizations in cooperative endeavors to influence everyday behaviour".[7]

As Bullock[8] argues, the cooperatives were part of the government apparatus from the start of the postwar era food system. They made up 88% of registered food dealers and were the market leader in the collection, storage, and transportation of food while also providing the main channel through which the government made crop payments to farmers. The relationship to the government was solidified in earnest when they bailed out many of the bankrupt local branches and mandated the reorganization and merging of the cooperatives through a Special Measures Law. The government had already set up a national level organization on paper, called the Nokyo Chuokai (Central Organization of Cooperatives) that locked in the restructuring of the cooperatives. Despite pressure from some members of the Liberal Democratic Party (which ruled Japan after the new Constitution and government were established) for a more open market for agriculture (and rice, the most important staple crop) cooperative leadership was strong in their vocal opposition to such plans. Furthermore, even by the early 1950s, the government had an interest in the success of the cooperatives and a government role in the food system, especially the distribution of rice because of its bailout of local organizations and substantial government investment in the restructuring and streamlining of a national level association.

As the postwar Occupation came to a conclusion and the Agricultural Land Law of 1952 established a reform of the landowning system that allowed more farmers to own their land, the role of the cooperatives did not diminish. The food shortages were over, but the cooperatives had already woven their way into the institutions that allowed the government to reach into villages by serving the state in a number of ways. The cooperatives were responsible for implementing both the rice reduction and crop diversification policies, as well as serving as a channel for the subsidies that the Ministry of Agriculture, Fisheries, and Forestry (MAFF) gives to local farmers in the form of loans, crop insurance, marketing, and promotion. The cooperatives function in the spaces between government and citizens thereby providing a vital linkage for both groups. Of course, the extent to which the cooperatives serve the interests of both of these groups is oftentimes debateable. Mulgan, in particular, argues that as *Nokyo* became the powerful actors that they are today, their interests increasingly diverge from those of its supposed constituents in order to maintain the existence of the current system.[9]

CENTRAL UNION OF AGRICULTURAL COOPERATIVES, JA ZENCHU

The organization at the helm of the cooperative movement today, JA Zenchu was incorporated in 1954, and its membership levels have at times accounted for 99% of all Japanese farmers. Bullock notes "(T)he organization was "grafted" onto local village structures, accounting for 99% membership as well as high group solidarity. A villager could be ostracized by his fellow farmers for refusing to sell his rice through the coops as late in time as 1969."[10] Its membership today is about 10 million, accounting for farmers and nonfarmer members (called associate members) who benefit by joining because of its vast list of services (including banking and loans, credit, funerals, and wedding halls). When JA Zenchu was set up, it was a semi-public organization, which was exempt from antimonopoly laws. This status gave JA Zenchu freedom to organize its vast array of activities and using its role as an implementing body for MAFF policies, the organization wielded considerable power. JA is integral to farmers, so much so that many farmers are dependent upon it, making entry into farming difficult unless one is a member. There is a tradition of farmers inheriting land from parents and other relatives, it is uncommon for people or companies with nonagricultural backgrounds to take up farming creating a system that is extremely closed to outsiders. Although recent reforms have intended to open up farming to agricultural corporations, as Chapter 3 notes, those outside of the existing structure are at an enormous disadvantage. The support of JA for these policies and the MAFFs most recent reforms are key to their success, it is involved extensively and throughout all levels of farming, both in establishing relationships between farmers and nonagricultural entities and through corporate membership.

The importance of JA in rice farming as well as other forms of farming today cannot be underestimated. The local cooperatives are involved and help farmers with every stage of farming, from providing them with seeds at lower prices, selling them fertilizers and pesticides, providing direction for planting, assisting with understanding changing growing conditions due to weather and other adverse conditions, such as disaster and pestilence, providing storage, drying, and processing services for rice, soybeans, and vegetables. Along with those services directly related to farming, JA also provides transportation services, an agricultural machinery section that assists with maintaining and fixing machinery, and providing advice through consultation and meetings of local farmers to coordinate how local cooperatives will meet prefectural goals and the implementation of national policy which is coordinated by JA with the MAFF. As its own webpage states, JA provides "backing to the central association of each prefecture, formulation of policies across the country, planning and development as well as information provision. We are coordinating policies and guidance at the national level with the MAFF and the JA Zen-Noh."[11] Along with activities that support farming, at the local level, the organization provides a range of activities including welfare services, a travel center, a gasoline station, a market, and childcare facilities. The list of local level facilities at local JA cooperatives is dizzying as Table 4.1, which shows those services available in Nanto City, Toyama prefecture where interviews for this research were conducted, illustrates.

The local branch office for this local JA publishes information about its branch, there are 2292 farm members and 1870 associate members. The total land area cultivated is 1646 ha and the focus of this region is *koshihikari* rice, as well as local vegetables, such as persimmon, onion, bell pepper, pumpkin, and ginger.[12] As one can observe by reading the list above, the services that are provided even to a small local cooperative are extensive and touch many aspects of a farmer's life. Specifically, the financial services that are used by JA members illustrate a deep involvement with the organization, that is, part of a lifestyle, rather than simply being an organization with a shallow commitment from members that utilize services infrequently. It is likely that a farmer in this area that is a member of JA, has a daily interaction with the organization, either through a visit to the bank, using the petrol station or shopping in one of its local A Coop stores. To say that the cooperative dominates country life is an understatement, it is a central mechanism for both the distribution and sales of agricultural products and financial services as well as a channel for distributing government monies and implementing national level policies as will be discussed in both this Chapter and the next.

Table 4.1   JA Nanto facilities list

Facility	Service offered
Head Office, Nanto City	General planning, auditing, financial consultation
Production Center	Sales and marketing, direct sales
Western Branch Office	Credit business, mutual aid
Shiinshin Office	Credit, ATM, counter service for savings and payment
Eastern Branch	Banking, mutual aid consultation
Iguchi Office	Banking, consultation
Gokayama Branch	Banking, consultation, sales and production
Shimpei Office	Banking, sales and production, mail and postal savings
Mutual Aid Center	Long and short term mutual aid
Western Country Elevator	Drying of rice, wheat and beans, production fermented compost
Kitano Country Elevator	Drying of rice, wheat and beans
Kamihei Vegetable Processing Plant	Manufacture and sale of vegetables
Agricultural Machine Section	Sales, repair of agricultural machinery
Divisional Field Office	Sales, reservation, delivery of rice, purchase of co-op products
Delivery Section	Sales, reservation, delivery of rice
Welfare Center	Nursing care, senior citizen welfare project
Travel Center	Planning domestic and overseas travel
Vehicle Center	Automobile repaid and inspection shop
Fuel Center	Full service petrol station
Car-Topia Inoguchi Village	Full service petrol station
Car-Topia Taira Village	Full service petrol station
Yotte Kare Johana	Direct sales of local agricultural products, souvenirs
Nanto Cooperative Store	Direct sales of local agricultural products and other goods

Despite their power however, the cooperatives are not viewed favoraby by many farmers, repeatedly during interviews, farmers were critical of the organization, saying that they existed for their own interests, not the interests of farmers. One farmer told me that nokyo "do nothing, they have not protected us."[13]

National Level Organization

At the national level, JA Zenchu (The Central Union of Agricultural Cooperatives) organizes and guides the activities of all of the cooperative related activities as well as providing oversight to its related agencies. JA Zenchu has 5.36 million nonfarmer members (or associate members) and 4.61 regular members, making its total membership 9.97 million. It has authority over 700 regional cooperatives and occupies the tallest building in downtown Tokyo, a source of pride for members illustrating the importance of agriculture to Japan. The primary activities of JA Zenchu are related to agricultural policy, including petitions, resolutions, and mass demonstrations. During election time, JA Zenchu is an important source of support for political candidates, its *koenkai,* or informal networking of citizens with donors and powerful supporters reaches deep into the rural countryside of Japan which accounts for a disproportionate share of representatives to Japan's House of Councillors, the lower and more important body in the legislative authorities. Most of Japan's legislators (at times as many as 45%)[14] have some connection to agricultural interests and agricultural policy issues.

The organization's related activities at the national level include its Norinchukin Bank that handles the credit and loan business including making loans to prefectural level cooperatives. The Bank was a target of a number of recent policy changes. JA's insurance company Zenkyoren is also part of the national level organization, providing life insurance as well as short-term insurance that covers profit loss from crop shortfalls and fire insurance. The final national level agency is JA Zen-Noh, which handles all of the marketing and distribution of rice as well as meat, fruit, and vegetables including fertilizer sales and organizes the prefectural level activities of all of the cooperatives. Zen-Noh is the largest feed grain importer in Japan, its share of the total compound feed production is 30% of the market.[15] JA Zen-Noh has a separate membership of over 5 million members in 709 prefectures.[16] The policy changes discussed below do not affect the activities of JA Zen-Noh, which is one of the weaknesses of the law.

Along with these national level bodies, JA also runs a tourist agency, a welfare organization, a set of newspaper publications and an educational publishing company. The bulk of the cooperatives activities are organized at the prefectural and local level. JA Zen-Noh organizes and oversees the activities of prefectural level JAs, and the Norinchukin Bank also has a prefectural level credit organization as does the welfare agency. The cooperative network in Japan is unlike any other because of the linkages that is has with the Liberal Democratic Party or LDP and the tribe of agricultural policymakers called norinzoku. As discussed in Chapter 2, this relationship is sometimes described as forming an iron triangle between bureaucrats, politicians and the cooperative organization. These groups have worked together in order to provide those living in rural communities and farmers with a panoply of benefits. As Mulgan describes, "(T)he agricultural cooperatives have been corporatized into the agricultural policymaking and implementation processes, with three-way policy negotiation and consultation taking the form of an institutionalized policy subgovernment (p. 263)."[17] This policy subgovernment continues to wield significant power and its budget allocation has not declined.

REFORM OF JA ZENCHU

In December of 2013, under Abe's Cabinet reform package the agricultural, fisheries, forestry, and regional vitalization plan was created. The stated goal of this plan was to double farmer's incomes and revitalize farming villages. The program was revised in 2014 to incorporate reform for JA. The two reforms outlined under the plan were to change the top-down structure of JA and to remove the national organization's (Zenchu) private interest in farming. This program underwent another change in 2016, which included the integrated reform of local level agricultural committees, agricultural production corporations, and agricultural cooperatives.[18]

On September 9th, 2015, the Act Partially Amending the Agricultural Cooperatives Act (Act No. 63 of 2015) was promulgated, it was enforced on April 1st. The act intended to reform agricultural cooperatives, with JA Zenchu as the target of much of its language, includes a lengthy list of revisions to existing law related to agricultural cooperatives. The list includes a partial revision of the Agricultural Cooperatives Act, a partial revision of the Agricultural and Fishery cooperatives Savings Insurance Act, a partial revision of the Act on Enhancement and Restructuring the Credit Business Conducted by Norinchukin Bank (an arm of JA), and Specified Agricultural and Fishery Cooperatives, etc. and the Repeal of the Agricultural Warehousing Business Act. For much of recent history, farmers had no other option when buying fertilizer and other products, and the JA charged them above market prices. Key areas of reform in the law include altering the right of JA Zenchu to supervise and audit regional and local cooperatives, a directive to become a general incorporated body rather than having semi-public status which prohibits it from antimonopoly laws (as previously mentioned) and restrictions on the cooperative's ability to compel members to use its services. Prime Minister Abe called the reforms "sweeping" in his April 29 speech to a Joint Session of Congress, including these reforms as part of Japan's "quantum leap."[19] The degree to which the reforms as enacted are sweeping, however, is questionable for a number of reasons discussed below.

The reform of JA's ability to audit and supervise cooperatives has only a superficial ability to alter the current way that the organization conducts itself. As Sugiura argues, the cooperative has already split its auditing division internally so that this function is "already almost completely independent."[20] This aspect of reform was intended to allow local and regional cooperatives to have more independence but they already have the ability to promote self-reform and have taken the initiative of entering into partnerships with a variety of distributors. A Japan Times editorial of February 2015 argues that the system which existed prior to reforms may have had little ability hamper local cooperatives initiatives and the administration has not shown how changing this auditing function would actually increase the income of cooperative members and assist in making farming more profitable and modernized (one of its expressed goals).[21]

With respect to the aspect of the law which requires JA to reform its organization, researchers have argued that this aspect of the law falls short of its goals as well. JA has already reformed its organization and the law is directed at the national level, while the extensive network of regional cooperatives or JA Zen-Noh is not forced to reform, so the legislation's ability to significantly alter this network is nil.

The aspect of the law aimed at changing JA Zenchu's legal status may also have little effect. The organization already works parallel with industry groups to include member's opinions, as organizations like it do, as Sugiura argues, these activities conform to many aspects of the reform already, lessening their impact significantly.[22] Moreover, JA Zenchu has already altered its organization (as mentioned earlier) and can separate its activities (many of which are already nearly independent of the national body) so that they conform to antimonopoly laws easily.

While the full effects of the new law won't be felt until 2019, when many of its provisions come into force, the organization has time to reorganize its activities in such a way as to prevent damage to its power and ability to lobby. As researchers have argued, the price adjustment system is not targeted or changed by the reform bill and this source of strength, that JA wields is its most significant.

THE 1952 SEED LAW AND ITS ABOLISHMENT

One of the current reforms that impact the role of JA in rice growing is the abolishment of the 1952 Seed Law, which occurred in summer of 2017. The law, affecting mainstay crops mandates the development and production of seeds (rice, wheat, soybeans, barley, and oats) by prefecture which are run by agricultural experiment stations. The law represents the legal basis for the experiment stations and its budget requests by prefectural governors to help farmers cover the cost of seeds. Under the law, prefectural governors designate the seeds recommended to local farmers and initiate a budget request which covered the production costs of seeds, allowing them to be sold to farmers at a low cost. The strict control over which seeds were used guaranteed that all mainstay crops were domestically grown, using domestically produced seeds. However, the abolishment of the law would allow the private sector, and large transnational corporations to enter into the seed business.

Critics argue that this might pave the way for Japan's seed technologies to become dominated by large Trans-National Corporations (TCCs) if they are able to develop and sell seeds. The sale of seeds would give the corporations inroads into farming in Japan and potentially the seeds could dominate Japanese farming in time. This is problematic for a number of reasons. Relying on seeds produced and sold by the private sector undermines Japanese food security. Also, the seeds sold by private companies are F1 hybrids whereas those developed and grown in Japan are filial 1 seeds. F1 hybrids do not reproduce seeds that can be used to grow next year's crop, so farmers must purchase new seeds every year. This can become very expensive and drive out smaller farmers, Also, the genetic resources and technologies regarding the staple crops grown in Japan are now currently managed by the agricultural experiment stations. If these stations no longer have a role in managing these important genetic resources, control may fall into the hands of private for-profit entities which manage them differently than domestic actors that do not have a profit motive. At the least, loss of management by the experiment stations undermines Japanese control over its own genetic seed resources.

THE POLICY IMPLEMENTATION ROLE OF JA ZENCHU

One of the most important roles for Japan's most important cooperatives, JA Zenchu has been as a policy implementer for the MAFF's programs regarding a number of activities. Foremost among these was the *gentan* (or set aside) system which issued payments to farmers to cut back rice production and either let rice fields law fallow or divert that land to growing other crops. The gentan system is most clearly and comprehensively explained by Mulgan, in her two books on agricultural policy in Japan as well as numerous articles.

The gentan is essentially a policy that pays farmers not to grow rice, which in the United States is also called the set-aside policy. The *gentan* or rice production adjustment system, as it was officially named began in 1970, and was administered and implemented by the MAFF's Agricultural Production Bureau until the reform of 2000 when the Food Agency took over this role and the program fell under the new Food Law system. Hayami and Godo[23] describe this system as the most consistent form of monopolistic control over rice distribution in the country. The gentan system was a national, top-down program which affected each prefecture equally until reforms were implemented in 2000. Meaning, that the requirement to cut back production was set at a nationwide level and implemented equally across the country to all prefectural farm households. The issue with this system was that good, efficient rice producing regions were required to cut back production to the same level as regions that were poor rice producing or even that only minimally produced rice. Also, those regions that produced rice in high demand such as koshihikari were forced to decrease planting. As Mulgan notes,

> Under such a uniform system, farmers found it difficult not to participate in the rice production adjustment programme. The equal allocation of gentan acreage to all farmers meant that if one farmer failed to play by the rules, all the others (in the local cooperative) would be penalised.[24]

In fact, many argue that rice farmers that were efficient were actually punished by the program which affected all prefectures and farmers equally preventing the concentration of rice production among efficient growers. The gentan system began to take a different form first in 1999 and 2000 when the Food Agency was abolished and the New Food Law brought about a set of reforms that shifted the responsibility for implementing policies that supported formers to a new policy bureau in the MAFF (the MAFF itself was also restricted).

New Food Law Reform

The reform of the gentan which took place in 2000, shifted responsibility for administering and implementing the program from the Agricultural Production Bureau in the MAFF, to the Food Agency. The rice production limit was included in the New Food Law, so the role of the Food Agency and the policy itself became law. Through the Food Agency, the government paid rice farmers to cut back on rice production and established strict control over the channels for distribution by requiring rice distributors to go through the process of being a designated seller through government channels. Mulgan argues that the effect of these reforms and goal of the changes was not to make rice production more efficient or to benefit farmers, rather the Food Agency under the MAFF's jurisdiction intended to bring all of the rice traded in Japan within its jurisdiction under the framework of the New Food Law and its plan for stabilizing the price of rice. This would consolidate control over the sale of all rice in Japan under the MAFF's control. Farmers were required to participate in the gentan, which was the main source of profit both for the JA cooperatives as well as the Food Agency. Some of the profit was paid back to farmers in the form of high prices but Godo and Hayami argue that the Food Agency and JA were the largest recipients of the profits.[25]

The supply and demand of rice and rice distribution therefore had an important function to play in this new system which had the goal of creating a stable supply and demand which were administered and controlled by the MAFF. Positioning itself this way, within the legal framework of the Food Law and rice distribution gave the MAFF extensive power and control over Japan's food supply and its most important crop, rice. Changes under the reform carried out in 2000 as part of the New Food Law were linked to raising food self-sufficiency as well as consolidating farmland and encouraging farming by full-time rice farmers. Although the payment program for farmers has persisted, it changed form in earnest with the set of reforms that began in 2000 and continued in 2005.

CORE FARMERS (NINAITE) PROGRAM

The first basic plan for agriculture under the new Food Law maintained direct payments to farmers but in 2005, the policy stressed the importance of identifying what it called "Core Farmers" who would be the foundation of a so-called stable and efficient system.

These core farmers are identified as the targets of policy support as well as community-based farming cooperatives who are also deemed potential core farmers. Core Farmers or *ninaite* (担い手) farmers whose income is comparable to workers in the nonfarm sector and who work equivalent hours. In other words, these are farmers who are efficient and likely make farming their full-time business. According to the MAFF, core farmers are also required to do farm management that is now or aims at becoming an efficient and stable farm in terms of production and income. Core farmers are identified by the MAFF according to two criteria:

1. Farmers must submit (5 or 10 year) plans to the local municipality to promote farm management, including farm size, income, family labor input and technological development. Local municipalities must then approve these plans.
2. Community based cooperatives that desire to establish themselves as *ninaite* must cultivate more than 20 hectares, and must aim to consolidate more than 2/3 of the land in the community. They are also required to have written rules of association, unify their financial accounts and have plans to become agricultural production corporations.

When farmers and local cooperatives meet these requirements, they receive preferential lending, tax incentives, and may be approved for certain land improvement and consolidation programs that make them eligible for grants as well. In order to quality for subsidies, farmers are also required to qualify as ninaite.

As a part of a shift from price-based policy to income policy, The Rice Farming Income Stabilization Programme (JRIS) was also introduced in 1998. JRIS compensates rice producers who participate in the planned marketing system for part of the loss of income when farm profits fall below the standard income that is calculated as the average, from profits of the three preceding years. Eligible producers have to fulfill the required diversion target of the year, enter into a contract with an agricultural cooperative and deposit a certain amount of money as a "limited withdrawal deposit" with the cooperatives. The JRIS program was revised in 2004 and the Core Farmer Management Support Programme (CFMS) was added to compensate the revenue loss exclusively for those core farmers who meet specific criteria. One of the major elements of this reform is the revision of the production adjustment policy, to be implemented in two stages.

The first stage of the reform was implemented in 2004, the production quota was allocated to each region based on the sales record of two preceding years, instead of specifying the area of diversion for each producer individually. The second stage of the reform, started in 2007, it allows farmers and farmer organizations to decide the distribution of the production quota. The role of government is foreseen to be limited to the provision of supply-demand information and to approve the production adjustment plan prepared by the producer organizations.

Currently, MAFF allocates the production quota to each prefecture according to the preceding four out of six years record of sales excluding the highest and lowest years. In order to provide an economic incentive to participate in the production adjustment program, MAFF provides subsidies to the producer's organizations that participate in the production adjustment program. Specifically, the diversion payment is allocated to the regional paddy farming associations, which are local JA cooperatives, according to the production quota. The regional associations are established at the level of the local municipality and prepare a regional paddy farming "vision" containing the future use plan of paddy field, the target level of planting and sales, and the specific usage of the diversion payment. In particular, the regional vision clarifies the list of core farmers based on the consensus between community members with the goal of allocating more resources to selected core farmers. In the previous scheme, a fixed amount of subsidy was paid to each farmer based on the diverted area of paddy. The new subsidies are paid to the regional associations who then allocate the diversion and subsidy among their members more flexibly. Regional associations are encouraged to distribute diversion payments to accelerate structural change voluntarily.

For example, the regional association in Chikusei city in Ibaraki pre-fecture succeeded in concentrating paddy use to core farmers (713 ha in 2003 to 1365 ha in 2006) by allocating an additional subsidy to core farmers and to landowners who rent out their land to core farmers.[26] The proportion of producer organizations that allocate diversion pay-ments focusing on the core farmers increased from 59 to 83% between 2004 and 2007.[27] In addition, the government provides short-term loans and subsidies for those who ship their crop separately to stockhold-ing in good crop years. The amount of diversion of paddy fields using the production adjustment program is calculated using a yield assump-tion which will be different from actual yield in any given year. Higher than anticipated production would have a negative impact on domestic prices if it were allowed into the domestic market and the loans to ship to stockholding acts as a buffer on supply in good years. In each of these programs a core role is given to the local JA cooperative as well as prefec-ture and the central union at the national level. The core farmer program is targeted at community or local level cooperatives who form a core farmer group. The income stabilization program or JRIS requires farm-ers to contract with a cooperative and deposit money with them, to be used as insurance against future loss of income. The cooperatives decide on production. The New subsidies and diversion payments are given to cooperatives which then dole out the money to core farmers as a subsidy and to local farmers who participate in diverting farming away from rice to other crops. The payment is not paid directly to farmers, so it con-forms to global trade mechanisms discussed in Chapter 3.

THE JAPAN DIRECT SUBSIDY PROGRAM: INCOME INSURANCE FOR FARMERS

Under Prime Minister's Abe's set of reforms designed to change agri-culture, likely the most drastic is the change in the direct payment sys-tem for farmers or gentan which has been undergoing changes as discussed above since the 1990s, when the language identifying core farmers was first introduced. The gentan takes on an altered form under the new agricultural policy which renames it as the Japan Direct Subsidy Program. This program, instead of compensating farmers for what they don't grow, or compensating farmers directly, functions as an income insurance program, or indirect payment against falling prices or decreased demand, to which farmers make payments toward. This sys-tem is very similar to farm legislation in both the European Union and

United Statesthat have adopted policies that are linked to loss of income. The income insurance will compensate farmers for income loss during a disaster or price reduction. The payment is based on the producer's total income, not by crop production which separates income insurance from rice growing. The plan is intended to mitigate against the impact of reduced income involving farmer contributions.

The MAFF describes it as a "safety net" focusing on the rice and field crops including wheat, barley, potatoes, and starches.[28] The plan is voluntary, and requires farmers to pay premiums that they would not be compensated for and would lose if a claim is not filed. Under the insurance program when sales fall below 90% of a farmer's average income (which the laws calls their standard income) over the past five years, a maximum of 90% of the difference between their standard income and the sales income can be compensated using a combination of two insurance programs.[29] One is a nonrefundable insurance program, covering 80% of the difference, the other is optional insurance which covers the remaining 10% of the difference. The new income insurance program replaces the previous Income Supported Direct Payment Program for farmers of 2010 which provided support for farmers in two ways. First, the program gave farmers a Fixed Price Direct Payment for producing rice when sold below cost and provided a Variable Price Direct Payment in the form of a subsidy when the cost of rice is sold below the average value. Both of these forms of direct payment is being phased out, the fixed direct payment was abolished in 2014 and the variable direct payment ends in 2018.[30] These are replaced by the 2007 Act to Stabilize Farmer Income which has two subsidies for farmers. The first of these is the income insurance program described above, the second is a supplement to assist with disadvantages in production conditions compared to other countries. This program is only available to core farmers or ninaite and is intended to promote farmland consolidation, efficient farm management by full-time farmers and to train a new generation of entrepreneurial farmers.

THE RICE DISTRIBUTION SYSTEM

One of the primary responsibilities and sources of power for JA has been as an approved primary collector, secondary collector, and national collector of rice. Under Japan's Food Control System which lasted from 1942 to 1995, the MAFF set up a strict rice distribution control that farmers used to sell their rice, please refer to Chart 4.1:

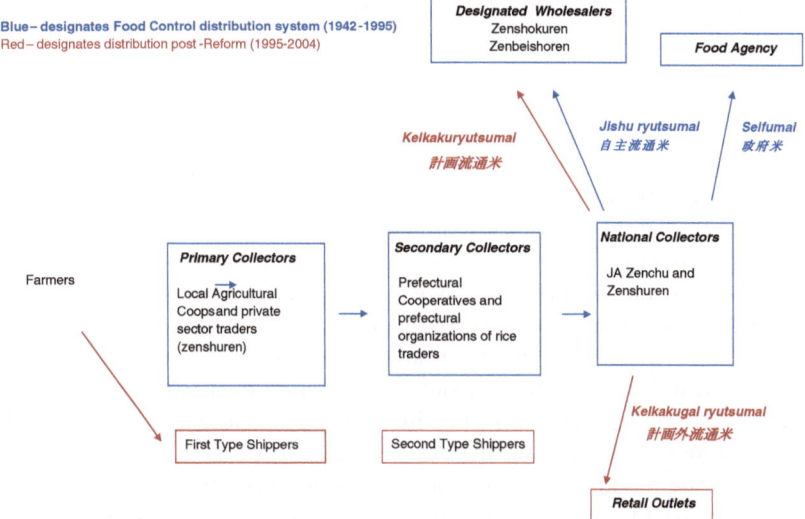

Chart 4.1   Japan's rice distribution system

Japan's Rice Distribution System. Selling rice was only allowed through these specific channels through designated rice collectors and wholesalers that it controlled. Farmers would first sell their rice either to the local agricultural cooperative (JA) or private sector traders (primary collectors) who then passed the rice on to the prefectural cooperative federations and/or prefectural organizations of rice traders (secondary collectors).

Then, these prefectural level secondary collectors passed the rice to national level collectors, either JA Zenchu or Zenshuren. Once in the hands of national level collectors, the rice would be sold either to the government or to approved wholesalers. In this manner, the system of rice distribution, storage, and collection was completely insulated from the market. In 1995, this system underwent a change when an additional channel for selling rice was opened up. This additional channel was called the non-orderly marketed rice market. The local, prefectural, and national level JA cooperatives played an important role in this process, which was coordinated through the MAFF. The MAFF and JA Zenchu worked together to keep players outside the rice distribution control system from having access, and this system was reinforced by all of the other activities of JA that kept the system running smoothly from the selling of rice seeds, the sale of fertilizer, insecticide, and through the use of the countless rice storage facilities that exist in every prefecture across the country. The Staple Food Law stipulated that production adjustment be used as a tool to balance the supply and demand of rice, which provided legal status to the production adjustment policy.

Rice marketing was partly liberalized with the introduction of a planned marketing chain composed of registered intermediaries in which producers would voluntarily participate. Direct government purchases in this system were limited to those made for stockpiling. Although the rice purchased through this program was set to reflect market demand, a price boundary was set to avoid volatile price fluctuations, which is a form of market manipulation. Producers were also free to market directly to consumers, which abolished the illegal sale of rice. Over time, the share of rice marketed outside of the planned system grew significantly over time making many of the later reforms a reflection of existing reality rather than a sweeping and substantial change.

Under the Cabinet reforms of 2004, the rice distribution system was changed removing all regulations on marketing and abolishing the former categories of government rice (seifumai) and voluntarily marketed rice (jishu ryutsumai) and abolishing the Food Agency. The government no longer designates mid-level purchases from rice from farmers and village cooperatives, rather the government requires that these rice purchasers are government-designated, which is a much less cumbersome program. The reform of rice distribution also removes the governments control over wholesalers and removes government channels of rice purchasing. To farmers, however, the system of rice distribution looks much the same, rice is still sold by local village cooperatives and may still be collected by prefectural JA and sold by JA Zen-Noh, it just no longer carries the seifumai label. The system of rice collection and transportation that existed for decades and is still very much in place, as one can clearly see by visiting any of the rice storehouses that are part of the JA rice distribution system.

The new program that began in 2013 and continued in 2014, replacing the gentan does not curtail production but it is implemented by JA Zenchu. Zenchu (through coordination with the MAFF) sets the target volume for rice production which can be produced in a year and then passes these allocations down to villages for implementation as part of the JRIS program described earlier. The Local Commission for the Promotion of Paddy Farming, prepares the plan for achieving the target. If the volume of rice production exceeds the target, excess rice is sold as staple food, at a lower price than cattle feed. Participation in the program qualifies participants to get favorable treatment in subsidy allocation and most of Japan's farmers (70%, according to Godo[31]) have decided to participate in this voluntary program. Along with the new program for rice farming distribution which liberalizes the marketing, trade and sale of rice, though to a far lesser degree than many, especially, in the METI would like because of the control that JA has over all of the infrastructure involved in rice distribution, the JA also implements the storage of rice.

Rice storage represents a significant part of rice distribution, during the rice shortage and price hike of 2007–2008, Japan kept its massive storehouses of rice from entering the market in order to ensure its own self-sufficiency but it had enough to successfully alleviate the price hike if it had chosen to release the rice sooner. The yearly stock required for storage under the new law is 1.5 million tons of combined government stocks and those that JA controls. The new program has a goal of diversifying the storage, transport, and storage of rice to future wholesalers and retailers. In practice, however, this is unlikely to occur, or if at all, very slowly because of the large investment required in rice storage and transportation. Local JA cooperatives store rice in what are called "country elevators" which exist in most villages across the country. Country elevators are used to dry, store, prepare and ship rice, wheat, barley, and soybeans. For example, when a farmer brings rice to the country elevator, it is weighed and dried, then it is stored in a large silo that is temperature controlled. This rice is pooled with the rice of other local farmers and then it is shipped to consumers, livestock producers for feed or is stored (Figs. 4.1 and 4.2).

Despite the sweeping reforms that Prime Minister Abe has talked up in speeches, numerous authors and scholars including those mentioned throughout the chapter, Mulgan foremost among them, argue that the reforms have not significantly changed the power of JA or the role that it plays in rice distribution, storage, and transportation. JA also continues to be very involved in the implementation of policy through the Farmland Banks and Core Farmer programs. It is likely that JA will

Photo 4.1   Country elevator, side view, Fukumitsu/Nanto, Toyama prefecture (Photo by author)

continue for many generations to come unless the bureaucracy which supports its activities undergoes ground-shifting changes, and this itself is unlikely given the conservative, slow process of change in Japanese politics. For farmers, however, this is not a win. The nokyo have maintained a system which is inefficient and unable to compete with mechanized farming and its larger economies of scale. The nokyo will preserve the current performative culture of rice growing to the detriment of entrepreneurial farmers and younger farmers who want to see Japan's rice growing lifestyle adapt, allowing for the industry to have a dynamic, vigorous presence. Currently, the industry exists because it is supported

Photo 4.2 Country elevator storage, Fukumitsu/Nanto, Toyama prefecture (Photo by author)

by an agricultural policy framework whose institutions exist largely to maintain their own power, (a comment repeated in my interviews with farmers[32]) they have kept rice growing unnecessarily antiquated, a policy step-child that demands large sums from the budget without a productive outcome. JA Zenchu and JA Zen-Noh are entrenched so deeply in the lives of rural communities and farmers that reform of the organization that would open rice growing to competition is unlikely. The current structure will persist until the policy actors and community changes substantially and calls are made to decrease agricultural spending.

Notes

1. JA Zenchu. 2017. "JA Multipurpose Cooperative and Activities." Available at www.zenchu-ja.or.jp/eng/multipurpose, 8/7/17. Accessed October 12, 2017.
2. Tsutsui, Masao. 2003. "The Impact of the Local Improvement Movement on Farmers and Rural Communities." In *Farmers and Village Life in Twentieth-century Japan*, eds. Waswo, Ann and Nishida Yoshiaki, 60–78. London: RoutledgeCurzon.
3. Dore, R. P. 1985. *Land Reform in Japan*, 2. New York: Schocken Books.
4. Ibid.
5. Dore, R. P. 1985. *Land Reform in Japan*, 295. New York: Schocken Books
6. Garon, Sheldon. 1997. *Molding Japanese Minds: The State in Everyday Life*. Princeton: Princeton University Press.
7. Ibid.
8. Bullock, R. 1997. "Nokyo. A Short Cultural History" (JPRI Working Paper 41), 2. Available at http://www.jpri.org/publications/workingpapers/wp41.html. Accessed October 12, 2017.
9. George-Mulgan, Aurelia. 2005. *Japan's Interventionist State: The Role of the MAFF*. London: RoutledgeCurzon.
10. Bullock, R. 1997. "Nokyo. A Short Cultural History" (JPRI Working Paper 41). Available at http://www.jpri.org/publications/workingpapers/wp41.html. Accessed October 12, 2017.
11. JA Zenchu. 2017. "JA 全中 について." Available at www.zenchu-ja.or.jp/about/outline. Accessed December 18, 2017.
12. JA Nanto. 2017. "JA なんとの概要." Available at www.ja-nanto.jp/company-outline. Accessed December 18, 2017.
13. Interviews. 2014. By the author with farmers and JA members in Hiroshima prefecture, August 1–30.
14. George-Mulgan, Aurelia. 2006. *Japan's Agricultural Policy Regime*. New York, NY: Routledge.
15. JA Zen-Noh. 2018. "Japan's Cooperatives." Available at www.zennoh.or.jp/english/cooperatives/japancooper.html. Accessed December 18, 2017.
16. Ibid.
17. George-Mulgan, Aurelia. 2005. "Where Tradition Meets Change: Japan's Agricultural Politics in Transition." *The Journal of Japanese Studies* 31 (Summer, 2): 261–298.
18. Fujibayashi Keiko. 2016. Japan Instituting Agricultural Reform Programs. *USDA Foreign Agricultural Service, Global Agricultural Information Network, Report JA 6065 12/22/2016.*

19. Prime Minister of Japan. 2015. "Toward an Alliance of Hope." Adderss to a Joint Session of Congress, April 29. https://japan.kantei.go.jp/97_abe/statement/201504/uscongress.html.
20. Sugiura, Nobuhiko. 2015. "Reforming of the Japan Agricultural Cooperatives." *The Japan News by the Yomiuri Shimbun*, 2. Available at http://www.yomiuri.co.jp/adv/chuo/dy/opinion/20150406.html. Accessed September 22, 2017.
21. *Japan Times*. 2017. "Sowing the Seeds for Lower Food Security?" Editorial. https://www.japantimes.co.jp/opinion/2017/04/13/editorials/sowing-seeds-lower-food-security/#.WhwYqktJneQ. Accessed November 27, 2017.
22. Sugiura, Nobuhiko. 2015. "Reforming of the Japan Agricultural Cooperatives." *The Japan News by the Yomiuri Shimbun*. Available at http://www.yomiuri.co.jp/adv/chuo/dy/opinion/20150406.html. Accessed September 22, 2017.
23. Hayami, Yujiro, and Yoshihisa Godo. 1997. "Economics and Politics of Rice Policy in Japan: A Perspective on the Uruguay Round." In *Regionalism Versus Multilateral Trade Arrangements, National Bureau of Economic Research*, 6, eds. Takatoshi Ito and Anne O. Kreuger, 371–404.
24. George-Mulgan, Aurelia. 2006. *Japan's Agricultural Policy Regime*. New York, NY: Routledge.
25. Hayami, Yujiro, and Yoshihisa Godo. 1997. "Economics and Politics of Rice Policy in Japan: A Perspective on the Uruguay Round." In *Regionalism Versus Multilateral Trade Arrangements, National Bureau of Economic Research*, 6, eds. Takatoshi Ito and Anne O. Kreuger, 371–404.
26. OECD. 2009. "Evaluation of Agricultural Policy Reforms in Japan." Paris, France: OECD Publishing. Available at https://www.oecd.org/japan/42791674.pdf. Accessed May 16, 2017.
27. Ibid.
28. Ministry of Agriculture, Fisheries and Forestry. 2017. *For Dissemination and Expansion of Good Agricultural Practices (GAP)*. Tokyo: Agricultural Production Bureau, MAFF.
29. Ibid.
30. Fujibayashi Keiko. 2017. *Japan Implements Agricultural Competitiveness Reinforcement Programs*. USDA Foreign Agricultural Service, Global Agricultural Information Network, Report JA 7029 June 26, 2017.
31. Godo, Yoshihisa. 2007. "The Puzzle of Small Farming in Japan." Asia Pacific Economic Papers, No. 365. Australia-Japan Resource Center, ANU College of Asia and the Pacific, Canberra, Australia.
32. Interviews. 2017. By the author with farmers and JA members in Toyama prefecture, July 27–August 1.

CHAPTER 5

Citizen Consumers: Cultural Protection and Japan's Food Movement

Japan has an active and diverse environmental movement that has a history of working to protect Japan's food supply. Women have been the initiators of this movement in large part, as they have acted first in order to protect their homes and their families from dangerous and polluting chemicals. Though this movement and their actions are often underplayed, the Non-GMO Movement in Japan as well as the environmental movement and consumer movement have had a number of meaningful victories in pressuring the Japanese government to protect Japan's food supply. Specifically, the environmental movements in the 1960s and 1970s were instrumental in pushing the LDP then to enact more strict laws to protect the environment. Moreover, the consumer and Non-GMO movements have had successes in setting standards for genetically modified food and country of origin labels. Historically, the origin of the environmental movement goes back to the 1960s and 1970s during the high phase of Japan's industrialization, when the danger of pollution had not yet been acknowledged. In Japan, consumers are more likely to be concerned about food safety, its origin and quality rather than price, and because of the history of Japan's consumer movements related to food safety, food issues have salience in the national media. This chapter details those organizations that are independent citizens groups, without a history of ties or a relationship to the Japanese government.

© The Author(s) 2019
N. L. Freiner, *Rice and Agricultural Policies in Japan*,
https://doi.org/10.1007/978-3-319-91430-5_5

BACKGROUND OF JAPAN'S ENVIRONMENT
AND CONSUMER'S MOVEMENT

The story of Japan's consumer movement including the Non-GMO Movement that formed out of the many consumer's clubs originating in the 1970s, has taken on renewed strength following the tsunami, flood, and nuclear meltdown (or triple disasters as they are called in Japan) that occurred on March 11, 2011. The following paragraphs provide an outline of the history of Japan's environmental movement to provide context and demonstrate the pattern that exists between the Japanese government and this particular group of civil society actors. The major features of this pattern include denial by government of the dangers of the pollutant posed, vocal citizen protest that is amplified by the media including alliances with international actors and finally reaction by government to address the problem (when the movement has used a strategy that has garnered the support of elite allies and the larger public). Furthermore, the successes gained by these movements are linked with the agricultural and fishing cooperatives in these regions.

Environmental disasters have had devastating consequences in modern Japanese history. These consequences included major pollution events such as the Itai outbreak (1972), the Minamata (1960–1974) outbreak and asthma in Yokkaichi city from 1970–1974.[1] Each of these events had associated diseases, which resulted from toxic substances traced to industrial pollution. The Itai itai byo ("it hurts") outbreak was caused by ingestion of cadmium traced to a metal refining company; Minamata disease was caused by poisoning from methyl mercury waste produced by a fertilizer company, and citizens in Yokkaichi city suffered from air pollution-induced asthma generated by the city's industrial complex.[2]

During Japan's heavy period of industrialization (from roughly 1950s-mid 1960s) many small, locally focused citizens' groups began to protest industrial pollution because of the severe health problems caused by the pollution mentioned in each of the three cases mentioned (mercury poisoning, cadmium poisoning, and air pollution). These groups fought denials of responsibility on the part of industry and unresponsiveness on the part of local governments. In the most comprehensive account of these early environmental disasters, Timothy George[3] writes, "the events of these years show a criminally irresponsible corporation achieving remarkable success at covering up its responsibility." The cover-up included the central government, that acted as an obstacle to cooperation among the medical and scientific communities and local government, that failed to protect citizens and even actively prevented good research by disbanding a group investigating pollution and lying about the causes of illness.[4] Schreurs[5] also notes the presence of reaction of the Liberal Democratic Party and the way that it acted as an obstacle to environmental progress.

An early example was in 1953 when the Ministry of Health and Welfare (MoHW) conducted a national survey of pollution that found that many Japanese were suffering from air, water, and noise pollution. On the basis of this survey and similar debates that were going on in the USA, the UK, and Germany, the MoHW formulated a bill to prevent contamination of the living environment. Other ministries, industry, and the LDP, however, opposed the bill. The unresponsiveness of both industry and government forced citizens to consider litigation, a political move that had been largely unused. The citizen's movement's use of the court system illustrates the seriousness of the government's unwillingness to protect its population from polluting industries, because it was a strategy that had not been used before.

The actions of government to deny the dangers that citizens were protesting, and the unwillingness of politicians including the most powerful political party (the LDP) to represent those concerns reflects a priority to industrialize at the expense of citizens' health. At the level of government, the two ministries responsible for these issues were at odds with one another. The tension between industrialization and environmental protection was being played out in the bureaucracy as the MoHW and the Ministry for International Trade and Industry (MITI) vied to protect their separate interests. Four large pollution cases brought to court by victims of the pollution events mentioned above forced major changes in a political system which had been closed up until that time. These court cases focused media attention on environmental problems, and it is widely acknowledged that the media helped turn national attention to the plight of victims of pollution in the 1960s and 1970s.[6]

POLITICAL PARTIES REPRESENTATION

Rising public interest in environmental issues gave opposition parties a chance to challenge the LDP in local elections, and eventually the LDP began to fear the salience of this issue on the national level. Environmental movements, which are traditionally localized phenomena, have received help in Japan from prefectural and district level labor union organizations, agricultural and fishing cooperatives along with other support groups as well as national and international assistance. Although opposition parties gave voice to the grievances of victims of environmental pollution, many of the most powerful movements used patterns of organization and networking separate from the established parties. Many times, opposition parties such as the Social Democratic Party and the Japan Communist Party mobilize these support groups in the early stages of environmental disputes.[7] For example, George[8] argues that because it was independent of the parties, the dominant stream of the Minamata movement was able to build a broad base of support that allowed the movement to achieve a favorable court decision with substantial public backing. Unfortunately, because of its independence (lack of connection with any formal political party) the Minamata movement did not continue after the 1973 court decision. The Minamata case presents a contrast to other protest movements which were swallowed up by the "national protest cartel, formed by a few political zaibatsu (mainly the JSP and JCP at the time), that monopolizes a large share of protest markets and eagerly swallows up new movements whenever possible."[9]

The literature on the relationship between environmental groups and opposition parties thus represents a contradiction: While opposition parties may help mobilize support for movements in early stages of environmental disputes, there is also a tendency for the parties to incorporate local movements into their political structure. This has the effect of dissolving the movement. This dynamic may be helpful to environmental movements who need party support to gain a political voice but it may also mean that some environmental groups do not become political actors in their own right and maintain the longevity necessary to build coalitions that are able to pressure higher level political actors. Although opposition parties represented the interests of some environmental protest movements, these movements were "swallowed up" as Steinhoff notes by opposition parties in order to strengthen the parties themselves, which tend to be weak political actors, especially at the national level.

The first significant win for citizens was the establishment of the Pollution Control Office within MITI in 1963 and in the MoHW in 1965. As a result of citizen initiatives, Japan went from having almost no environmental policies to having among the strongest environmental laws in the world.[10] For example, the Environmental Pollution Control Act and the Nature Conservation Law were enacted and Special Standing Committees for Industrial Pollution were established in both houses of the diet and an Advisory Commission on Environmental Pollution was created under the MoHW's jurisdiction. The MoHW was still at odds with MITI, which argued that environmental legislation should not hamper the growth of Japan's industrial sector, and as a result the 1967 Basic Law was adopted without the inclusion of strict liability for polluting industries advocated for by the MoHW. Additionally, in 1968 the Air Pollution Control Law was passed, setting standards for sulfur dioxide emissions, following a Cabinet ordinance on ambient air quality standards for SO_2. Also, in 1968, the PCB pollution outbreaks in Kanemi led to stronger demands from activists for significant change, alongside this development activists in the US started to win victories that gave legitimacy to environmental demands worldwide.

By far the most important legislative victory occurred in 1970, when the diet held a special session on pollution, enacting 14 antipollution laws and establishing a Headquarters for Pollution Countermeasures. Furthermore, in 1971 the Environment Agency was established, the National Park and Air Pollution Divisions were taken from the MoHW and put in the new agency and four agency bureaus were created: air quality, water quality, nature conservation, and planning and coordination[11] (Schreurs 2002). The environmentalist movement in Japan grew out of citizen response to the pollution of coastal areas and marine life as a consequence of industrial practices. While the environmental movement in Japan was health driven, it was successful in accomplishing major changes in the way the Japanese government dealt with the environment. However, by the mid-1980s there was a decline in the environmental movement in Japan. The reason for this decline can be attributed partially to the positive measures taken by the Japanese government addressing many of the worst environmental problems and the fact that the quality of the environment had taken a noticeable turn for the better.[12] Alongside the success of government policies, Japan enjoyed a time of economic success during the 1980s and Japanese citizens were reluctant to criticize their government or its environmental policies while enjoying the benefits of such prosperity. Without pressure from citizens groups the Environment Agency was unsupported and MITI returned to protecting industrial interests; consequently Japan's economy took off during this period and the country enjoyed the highest per capita GNP in the world. Despite the decline in environmental activism through the mid-1980s, there was a return to concern for environmental policy in the late 1980s that continued through the early 1990s[13] (Schreurs 2002). In the most recent election cycles, discussed in Chapter 2, the dominance of the LDP has strengthened Japan's centralized system, and Hasegawa argues that the result has been ineffectual deliberative bodies in the diet and other councils of review. Given this context, the contributions made by environmental movements to government decisions have been substantial because of their ability to influence public opinion and the policy process to achieve policy change. The vocal protest movement that emerged during the Fukushima Daiichi protest for example, which will be discussed in detail later in this chapter is one example of a citizens group that has garnered media attention, bringing to light problems of radiation in the food supply and the government has responded to these problems and the voices of its citizens.

OVERSEAS DEVELOPMENT AND GAIATSU (外圧)

In the 1980s several groups in Japan began to address international environmental concerns and a fledgling national presence of environmental organizations emerged.[14] During this time frame consensus grew regarding the impact of Japan's postwar economic development on the global environment. This return to a concern with environmental matters reflected a growing international awareness of environmental issues as well as criticism of Japan's overseas development policies. As a major importer of food, energy, and resources, Japan is responsible for the depletion of natural resources and pollution both at home and abroad. One article from this period calls Japan "an "eco-outlaw, a whale killing, forest-stripping bogeyman on the environmental stage."[15] Japanese consumption accounted for 40% of the world's imports of wood, mostly from tropical, undeveloped countries in the late 1980s-early 1990s; Japanese fishing vessels engaged in the practice of drift netting and whaling, killing thousands of porpoises, seals, and dolphins and endangered species of whales. In 1989, the United Nations Environmental Program ranked the Japanese the lowest in overall environmental concern and awareness.[16] In 1992, when Japan hosted the International Convention on Trade in Endangered Species, activists were "aghast at the Japanese fishing industry's fierce lobbying campaign against a proposed ban on imports of West Atlantic bluefin tuna."[17]

Along with destructive environmental practices due to consumption of the world's natural resources, Japan's economic assistance policies also had powerful environmental consequences. During the early 1990s, Japan was one of the largest contributors to developing countries with development assistance. As a country with such economic clout in the world, Japan was able to substantially alter the global environment. Much of Japanese foreign aid went to large development projects such as manufacturing plants and mines, which have high environmental costs. For example, Japanese Official Development Assistance (ODA) helped to fund, in part, the Narmada dam project in India which was protested by local citizens and international environmental groups. In the early 1990s development assistance was the fastest growing item on Japan's national budget and, "often those aid packages came with a disturbing quid pro quo: developing countries use the funding for projects that yield (and deplete) resources needed by the Japanese."[18]

Japan's development policies at home and abroad generated international criticism which affected the government's posture on environmental affairs, especially at the global level (Mason 1999). During the run-up to Rio (the 1992 United Nations Conference on Environment and Development), as well as at the conference itself, Japan's Liberal Democratic government sought to counter negative opinion and position the nation as a model global environmental citizen. "In contrast to the situation in the early 1970s, these recent initiatives were largely government-inspired; proactive rather than reactive...In December 1997 Japan hosted the United Nations Framework Convention on Climate Change (COP 3) in Kyoto and, again, the government sought to exercise environmental leadership by brokering an agreement."[19]

Japan's government has been responsive to international pressure in some issue arenas when there is particular concern about the way the world views Japan. Gaiatsu 外圧 (foreign pressure) has made the Japanese government more proactive on environmental issues to some degree, although this reflects poorly on the government responding to citizen's concerns proactively. Mason notes that as one of the world's leading ODA donors, it is important that Japan's development-related activities be viewed favorably. "Gaiatsu was employed early on in the modern environmental movement when Minamata victims embarrassed the government by their appearance at the 1972 Stockholm Conference on the environment."[20]

In 1990, Japan also canceled ODA credits for the Narmada dam project in India in light of highly visible international pressure and criticism which forced the World Bank to withdraw support from the same project. This change in direction from the policy retreat noted in the 1980s stems from an increase in international awareness of environmental problems coupled with the recognition that the development policies of industrialized nations have devastating consequences on environmental degradation in the developing world. At the sixth Conference of the Parties to the United Nations Framework Convention on Climate Change (COP6) held in 2000, Japan earned the shameful moniker of "Fossil of the Day" for two days in a row because of its negligible commitment to global warming when these activities are seen as adverse to Japanese national interest.[21] In order to conform to international environmental policies, in 1993 Japan adopted the Basic Law, which integrates sustainability (the main theme of the Rio Summit) as its mission and includes recommendations for a basic environment plan, environmental impact assessment, economic measures, environmental education, promotion of NGO activities and promotion of science and technology.[22]

GENETICALLY MODIFIED ORGANISMS IN JAPAN

Genetically modified organisms or GMOs did not emerge as a concern for most Japanese until the mid-1980s, when the Japanese public became to raise the matter in public debate. Up until that time Japan had no history of growing GMOs domestically. Citizens groups, such as the Seikatsu Club Seikyo, a consumers organization focused on issues of home economics and food safety with broad membership across Japan as well as other consumer groups began the public outcry against GMOs in Japan which intensified in 1998 when the sale of genetically modified Hawaiian grown papaya called Rainbow Papaya first attempted to be sold in Japan.

The papaya was one of the first foods to be introduced in Japan and it sparked a controversy about GMOs, 1999 was a year of intense debate in Japan and the public outcry against GMOs led the country to adopt a food labeling requirement for all genetically modified foods that is one of the earliest food labeling laws in the world targeting GMOs. The law, housed in the MAFF requires that all genetically modified foods be labeled accordingly. Along with the food labeling requirement, Japanese law also set regulations for government experiments. No domestic planting of GMOs is allowed and research by private corporations is also prohibited. Along with the law targeting the labeling of GMOs the MoHW also set standards in the Food Sanitation Law that requires GMOs to be labeled.

The problems with GMOs are mainly related to the contamination of organic agriculture and foods by GMO crops that can occur along all levels of the food chain. Contamination refers primarily to horizontal gene transfers or the transfer of genetic material between sexually unrelated and incompatible organisms which is more likely in genetically modified plants that can spread into organisms that wouldn't normally have them. Also cross-pollination is a concern, cross pollination occurs within related organisms, generally either across seeds or crops. Along with the contamination across plant breeds, the creation of pesticide-resistant pests and superweeds resistant to weed killers like roundup are potential hazards. When horizontal gene transfers take place between the crops and weeds, herbicide-resistant weeds are created. Also multiple resistance or gene stacking can happen when weeds or volunteers are contaminated with herbicide-resistant genes for several different herbicides. These new genetically modified organisms threaten existing weed strains and crop biodiversity which is a key concern of environmental scientists. The risk to crop biodiversity is particularly disturbing because wild crops and indigenous landraces (groups of plants that are especially vulnerable to GMO contamination) are valuable reservoirs of genetic information which may hold the keys to disease prevention and potential cures. Moreover, established seed stocks may also be threatened as well, these are seeds that have been the result of hundreds and even thousands of years of evolution and whose existence has been the result of natural selection.

The key citizen group actors in Japan that are working on GMOs is a short list, topping the list is the Citizens Union of Japan or (CUJ) which is led by director Martin Frid. The CUJ has worked on GMO labeling since they first raised the issue in the early 1990s after the Japanese government had approved the domestic sale of imported GM soybeans, corn, and other grains. The CUJ started the campaign to demand mandatory labeling and an organization committed solely to the GMO issue was created by the Non-GMO campaign for CUJ. This Tokyo based Non-GMO campaign (www.gmo-iranai.org) has the goal of a GMO free Japan. Along with CUJ and the Non-GMO campaign, the Seikatsu Club Seikyo and the Green Coop Consumers Co-operative are all committed to strengthening Japanese regulations on GMOs and protecting Japan's food safety. The legislative successes they have achieved are notable, especially given the Japanese governments history of denial and lack of transparency about safety concerns that affect Japanese families. The reasons for success can be attributed to the widespread support of these groups and the public concern for GMO food in the 1990s. For example, the groups were able to obtain over half a million signatures on a petition against Monsanto's trials of genetically modified rice in 2001.

The media covers food as a key issue in Japan, and the reporting of food issues allows the public to participate and act on food matters. The concerns of Japanese citizens mirror those of the larger public, whose aversion to GMOs is a phenomenon that exists around the world.[23] Japan plays an important role in biotechnology because of the large share of imported foods in its domestic market.[24] Prompted by citizens groups, the government created a number of laws and institutions with which to deal with the GMO issue. Along with the public support of the work of these groups, the collaboration between consumers groups and farmers, including farmer's cooperatives have contributed to the success of the movement. As noted, "In Japan, the cooperative organization is well established and strong, so we can draw on its power."[25] The real risks of GMOs are numerous, but they most pressing is the contamination of the agriculture and food chain at all levels. The vulnerability of the food chain has been illustrated in a number of cases involving contamination by GMOs. One of the most resilient strands of a genetically modified crop is Bt Canola. Bt stands for Bacillus thuringiensis, the name for a naturally occurring mold that is pest resistant and which has been genetically grafted onto a number of crops to make them pest resistant as well. Bt crops are now commonly used in the United States where large conglomerates like Bayer and Monsanto have developed varieties of pest resistance corn, maize and canola all using this gene. The risks of using such plants mainly involve genetic transfer and cross-pollination which threatens non-GMO and organic agriculture as well as native species which threatens biodiversity, key concerns for biologists and plant scientists. The impact of biodiversity is especially problematic in the case of wild, native and indigenous varieties of plants whose genetic histories contain valuable lessons for scientists and are valuable resources for new and useful genetic traits.

JA Zenchu's Efforts on GMO Issue

Working with consumers groups, such as the Seikatsu Club Seikyo and Rengo, JA Zen-Noh extended a five-year contract with seed developer Hi-Bred International (the seed subsidiary of chemical giant Dupont) to develop non-GMO corn seeds for American farmers who export corn to JA Zen-Noh. Zen-Noh imported GM corn for beverage and seek companies, but in order to promote the demand for GMO crops at home "established a council to promote non-GM corn at home and abroad and call on farmers to continue production."[26] JA Zen-Noh has committed to working with companies to secure a stable supply of non-GM corn for Japan's nonfood market. Currently, GMOs are included in all food labels and are not produced domestically. However, GMOs still account for about 40% of imports of corn, soybeans and other grains that are used for feed and the production of additives such as cornstarch.

JA Zenchu was also a signatory to the Joint Declaration on Food Sovereignty by farmer's organizations from developed and developing countries that took a common position on WTO negotiations in agriculture in 2005. The statement issued by farmers in Japan (as one of the reasons for signing the declaration) included the right for family farmers and peasants to practice sustainable farming and for governments to act to ensure food safety that addresses their concerns of their citizens, noting the concern with GMO food. JA has been an active and vocal defender of food sovereignty as well as supporting the movement against GMO foods as part of its goal to protect family farming in Japan, sustainability, and the livelihoods of its member farmers.

In Japan, the political power of farmers and fishermen's organizations is linked to what Putnam[27] and others call social capital is measured by the membership and participation of citizens in community organizations, including fraternal, community, benevolent associations and groups that make up political and public life. The concerns of fisherman and farmers have had a direct effect on the siting decisions of government officials. "Nuclear blight" is the fear or rumor that radioactivity contaminates produce and harvests, prompting fears of consumers resulting in the loss of sales; it is a fear that both groups take seriously. The strengthening of civil society is related to the presence of the farmer's and fishermen's associations in Minimata (Kumamoto prefecture), Toyama (Toyama prefecture) and Yokkaichi city (Mie prefecture) that protested industrial pollution. To this day, nuclear power plants have not been cited in either Kumamoto, Toyama, or Mie prefectures, because the government has avoided those areas where citizen protest is likely.

MAFF REGULATIONS ON GMOS

With regard to GMO's the MAFF plays a key role in the approval process for GMO crops. The Japanese legal regulations on GMOs are based on the Cartagena Protocol on Biosafety to the Convention on Biological Diversity that was opened for signature in 1992 and entered into force in December 1993 for the Conference of Parties to the Rio Summit, the agreement entered into force in Japan in 2004. The Protocol reaffirms the precautionary Principle of the Rio Declaration and maintains the trade and the environmental are not exclusive but are mutually supportive with regard to achieving sustainable development. The protocol defines biotechnology, including GMO's (sometimes referred to as living modified organisms, LMOs).[28]

Public aversion to GMOs around the world along with the high percentage of imported foods in Japan's domestic market and the public outcry and protest involving GMO foods has led the MAFF to advocate for its agricultural constituency, pushing for regulations that will advantage Japan's domestic producers. Japan's domestic legal regime on GMOs responds to citizens concerns about GMOs and a number of recent events that have gained media attention on the issue. Foremost among these are the controversies around Rainbow Papaya, Star-link Corn, and Golden Rice. Rainbow Papaya was one of the first genetically modified foods to be marketed in Japan, it first became available in 1998, after a Hawai'an scientist developed the papaya after the native species were attacked by ringworm virus essentially ending the ability for Hawaiian farmers to grow papaya.

When Japanese consumers became aware that the papaya was genetically altered, citizen's groups protested and the public and media led an outcry against the genetically modified fruit. Japan had not yet created its own regime for implementing the Cartagena Protocol but the government responded by suspending further shipments of the plant. Starlink corn is a GMO that was found in the US food supply in September 2000 and caused a nationwide recall in the US of more than 300 corn-based foods. At the time, it was approved only as a feed due to concerns that it might cause allergic reactions in humans. Japan's import inspectors (part of the MAFF) detected star-link corn in a US corn shipment in a vessel docked at Nagoya harbor, the imported corn was intended for use as foodstuff (cornmeal). Star-link was not even approved as an animal feed in Japan, although it was authorized for such use in the United States.[29] The incident set off a number of protests including one outside the US embassy including representatives from the CUJ, No GMO and JA Zenchu. Recently, Monsanto has successfully developed a genetically modified rice, called Golden Rice which according to the company alleviates the cedar pollen allergy. Experiments growing this strain of rice are being conducted by the National Institute of Agricultural Science in Japan or NIAS (Table 5.1).

The leading piece of international law on GMOs is the Cartagena Protocol, Japan's legal regime is based upon this document. The Protocol was opened for signature at the Rio Convention on Biological Diversity in 1992 and entered into force in December 1993. It reaffirms the precautionary principle of the Rio declaration and views trade and the environment as mutually supportive with the view toward sustainable development. The sustainable development paradigm values indigenous knowledge and biodiversity. According to the Cartagena Protocol, there is a distinct set of regulations for products intended for consumption (Type 1) and feed versus those for research and laboratories (Type 2). Products that will have Type 1 usage must only be approved when they are judged to do no harm to the biological diversity of the planet. Type 2 uses must ensure proper containment, handling, and usage so that the organism is not dispersed into the environment.

The MAFF is part of the process involving all Type 1 usage for crops, live animal vaccines and any other use that involves agriculture including crops, dairy, and livestock. The MAFF is also involved in Type 2 uses involving any GMO used as feed. The process for Type 1 usage is lengthy, starting with a required literature review and testing before an

Table 5.1 List of laws

Law	Ministry
Cartegena protocol (2004)	MOE, MHLW, MAFF
Food Sanitation Law (2001)	MHLW
Food Safety Basic Law (2003)	MHLW, MAFF
Food Labeling Act (2015)	MHLW, MAFF, MOE

application is even submitted. Type 1 usage refers to any instance where the GMO may be released into the environment, covering approvals for GMO crops, live vaccines, and open field-testing of GMOs (before they can be approved to be grown as crops). These tests are to be conducted in a similar environment as the one that will be used in Japan starting in the laboratory and then if it is deemed to be safe, open field tests. After these tests are conducted, the applicant prepares an Assessment of Adverse Effect on Biological Diversity and the other components of the Application (Emergency Measures and a Monitoring Plan) see Chart 5.1: Pathway for Type 1 GMO Approval, Ministry for Agriculture, Fisheries and Forestry (MAFF) below for an outline of the full process.

Prior to submitting the application, the Ministry encourages applications to consult with the Plant Products Safety Division, that is part of the Food Safety and Consumer Affairs Bureau of the MAFF. The applicant will eventually be filed with this division, one it is completed. After the application is filed, the MAFF conducts a hearing by experts, also called a "Review Board" which examines the application in accordance with regulations set forward by ministerial ordinance. Ministerial regulations act as method of implementing national law in Japan, in this case they are set forth to implement the Act on the Conservation and Sustainable Use of Biological Diversity.

After the hearing by experts, if the genetically modified organism is deemed to have no risk of harming biological diversity, then the public is notified and public comment is invited. After this comment period passes, and there is no significant question raised about the organism it is deemed to pose no risk, it is approved. After approval, the person or corporation that submitted the application must create an oversight committee and appoint a Management Representative and Management Supervisor in order to continue to assess the risk to biological diversity and to maintain appropriate cultivation, storage, transportation, disposal and acts incidental to them. The MAFF states in its guidelines for application that the approval process (after application) takes approximately six months. If the GMO is not deemed safe for field-testing because it poses a possible risk to biological diversity (following the precautionary principle), the application will be denied. If there is a lack of information, the applicant may be requested to revise and resubmit the application again. At each step along the way, the Ministry of Environment is involved and has final approval according to its own ministerial ordinances (Table 5.2).

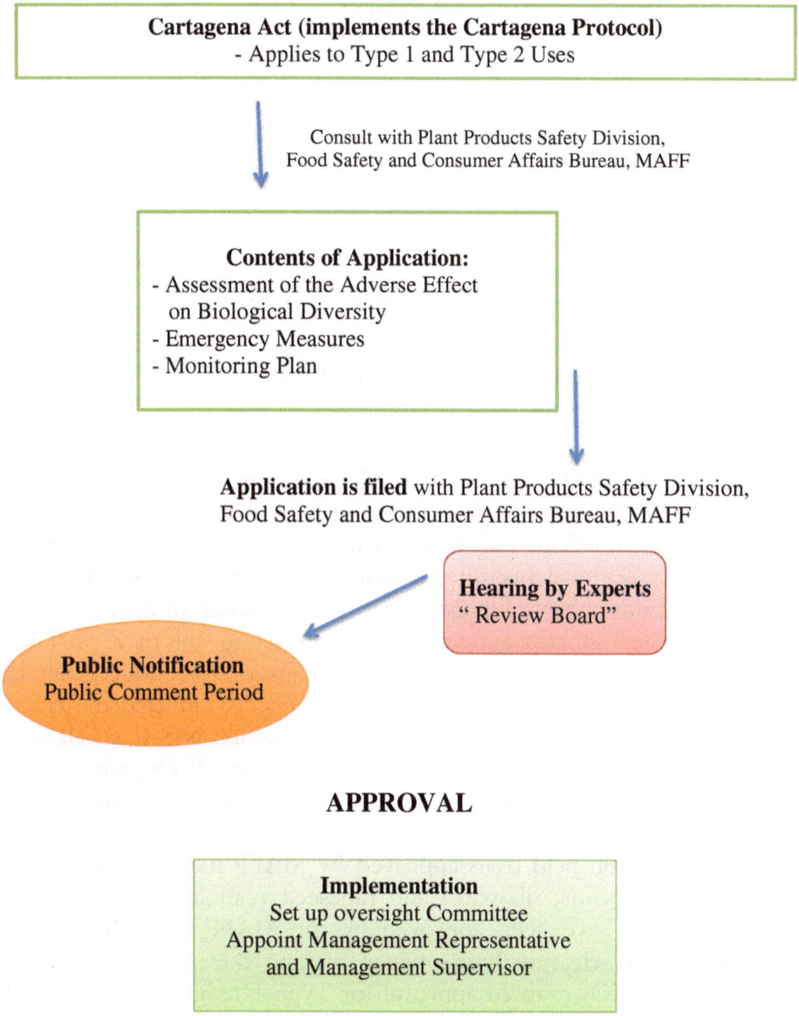

Cartagena Act (implements the Cartagena Protocol)
- Applies to Type 1 and Type 2 Uses

Consult with Plant Products Safety Division,
Food Safety and Consumer Affairs Bureau, MAFF

Contents of Application:
- Assessment of the Adverse Effect
  on Biological Diversity
- Emergency Measures
- Monitoring Plan

Application is filed with Plant Products Safety Division,
Food Safety and Consumer Affairs Bureau, MAFF

Hearing by Experts
" Review Board"

Public Notification
Public Comment Period

APPROVAL

Implementation
Set up oversight Committee
Appoint Management Representative
and Management Supervisor

Chart 5.1   Pathway for Type 1 GMO approval, MAFF

Table 5.2   Ministerial ordinances on GMOs

Ordinance	Ministry
Standards for the safety assessment of GMO	MOE, MHLW, MAFF
Policies regarding safety assessment of stacked varieties of GMOs	MOE, MHLW
Starnds for the safety assessment of food additives using GMOs	MHLW
Policies regarding safety assessment of non-protein food additives	MHLW
Ordinance to designate measures to prevent dispersal of GMOs, industrial use	MOE, MAFF
Ordinance to designate measures to prevent dispersal of GMOs, R&D	MOE, MAFF
Act on Food and Agricultural Materials Inspection Center	MAFF
Act on Safety Assurance and Quality Control of Feed (2007)	MAFF

In 2015, the MAFF approved of 22 applicants for open field-testing, and for the first time these tests included rice. Prior to 2015, only two instances of GMO field tests were allowed, these occurred in 1999 and 2006. Japan's National Institute of Biological Sciences (NISA) submitted all of applications approved in 2015, for testing a strain of disease-resistant rice that may also enhance agricultural productivity. To this date, no foreign or domestic corporation has been granted permission to conduct open field tests for rice. All of the approved trials have been conducted by national research organizations (including the aforementioned NISA as well as the National Agricultural and Biological Oriented Research Organization and the National Agricultural and Food Research Organization) and Tohoku University.

The other open field trials approved by MAFF include tests for cotton, maize, soybeans, silkworm and rapeseed (canola). A total of 314 lines of GMOs have been approved by the MAFF since 2004, when Japan's biosafety clearing house began making the information public. Most of the GMOs granted approval for Type 1 usage have been applications for the transportation and storage of organisms, and for the use of GMOs as food or feed. Rice, carnations and a small number of other GMO plant varieties (bentgrass for example) have been approved, either for cultivation or open field-testing. The only plant approved for commercial cultivation currently is a blue carnation that has been judged to pose no risk to biological diversity. When considering applications for Type 1 usage, three primary considerations are taken into consideration:

(1) the impact of the organism on competition between native Japanese plant species, (2) the potential impact of the GMO hybridizing with native wild species, and (3) the potential for the GMO to produce harmful substances that will impact the environment. There are no Japanese plants that will crossbreed with rice, maize or canola but there are native species that will hybridize with soybeans, and roses so these particular groups are examined more closely.

Along with acting as the government agency in charge of implementing laws regarding GMOs for use in open fields, the MAFF also oversees applications for live vaccines and GMOs that are used as feed (which may then also turn up in the food supply). Furthermore, the MAFF is responsible for inspecting shipments of food and feed as they enter the country for possible unlawful entry of genetically modified feed and plants. The MAFF has yearly conducted inspects of ports of entry for the dispersal of maize, rapeseed, and soybeans and has found no cross-breeding with domestic plants. However, some citizens groups dispute these results. The CUJ and NoGMO! Campaign have both conducted independent monitoring of canola and found dispersal beyond the ports.

The movement against GMOs in Japan that was recognized by the media and academics was responding to the controversies in the late 1990s described earlier with Rainbow papaya and Starlink corn. The most important of these is the Consumer's Union of Japan, which has conducted its own independent tests and one of its leaders then created a new organization focused solely on GMOs, the NoGMO! Campaign (or we don't need GMOs campaign in Japanese). Both organizations have led protests against GMOs, rallied the public cry to protect Japan's food supply and organized the signing of petitions to halt GMO trials by large corporations (the most recent was against genetically modified rice, which is currently being developed in the USA). The media attention and public support of these groups, along with the support of powerful allies such as the farmer's cooperatives has allowed the campaign against GMOs in Japan to sustain their demands with government bureaucrats and maintain the pressure on government accountability. These groups have been vital in sharing information with consumers both in Japan and abroad because they publish information in both English and Japanese. Along with these two movements, the Citizen's Bio-Technology Information Center is a website which publishes information in English in Japanese that would otherwise likely get swept aside because it does not make headlines. The website publishes a

Table 5.3   List of citizens food safety NPOs

Organization	Issues
Consumer's Union of Japan	Consumer's rights, food safety, GMOs
Citizens Bio-technology Information Center	Food safety
No GMO! Campaign	Food safety, GMOs
Pacific Asia Resource Center	Food safety, consumers rights
Citizen's Network for Sustainable Agriculture	Sustainable farming, organic farming, GMOs
Seikatsu Club Seikyo	Consumer's rights, food safety

monthly list of stories related to biotechnology and GMO's in Japan for its Food Safety Citizen's Watch. The citizen's movements on food issues and GMOs is strong in Japan today with a variety of organizations acting at the national level, Table 5.3 is a list of the more prominent of these groups.

One of the issues motivating these organizations has been food labeling, which reached a high water mark of activity in 1999, a year of intense disputes between government and citizens groups protesting against the import of genetically modified soybeans, corn, maize that entered the country with government approval in 1996. That same year, the Ministry of Health, Labor and Welfare's advisory board on food safety conducted a special session on biotechnology for the first time in Japanese history to discuss the GM crops being imported. Between 2002–2007, Monsanto began conducting open field trials with GM rice varieties, but local disputes against such testing eventually brought the case to the Tokyo High court which rejected the case stating that there would be no substantive ecological damage. Following the decision, consumers boycotted products of the companies importing GM food and conducting the tests (Mitsui Chemical, Kagome, and Kirin). Moreover, the local farmer's cooperative sued the experiment station doing the open field tests. Consumer's and cooperatives allied to petition the companies against GMOs, the sheet with 350,000 signatures was presented to the prefectural governor in Hokkaido. Under this pressure, the corporations stopped the experiments. In 2001, the law requiring mandatory labeling of GM foods entered into force. The law requires that any food, feed, crop, vaccine, plant that is genetically modified must follow the process for approval illustrated in Chart 5.1.

However, one weakness of the food labeling law is related to food additives. The current law required that when the final products contain the same DNA as the raw material it is not required to be labeled as genetically modified (this rule differs from European standards which requires traceability). Also, the Japanese law has a 5% threshold regarding the presence of genetically modified material in food. If the genetic modification represents less than 5% of the total weight of the final product then labeling as a GMO is not required. As a result of the lack of a traceability requirement and the 5% threshold, consumers may purchase genetically modified foods without their knowledge. Non-GMO foods are labeled voluntarily, and must be segregated in the supply chain. In 2001, when GMO labeling was first introduced, citizens groups criticized the voluntarily labeling aspect of the law, but increasingly many Japanese growers are adopting the labeling because it offers them an advantage with consumers. Citizen's groups feel that these efforts are not enough, however they show at least some measure of government response. The degree to which this response is motivated by concern about its citizens rather than control over another area of food policy, for which it may request funding is beyond the scope of this chapter. Given the framework of agricultural policy in Japan and the way the MAFF formulates its policies, the efforts of citizens with regard to GMOs has the potential to be co-opted by the MAFF in order to elevate the cultural importance of Japanese rice. Certainly these policies can help the government make claims about the purity and unique nature of rice growing in Japan, these claims elevate the ministry's importance. With the government's promotion of small farmers and washoku, the ability for consumers to distinguish between foods is playing a higher role in the Japanese food market.

THE TRIPLE DISASTERS AND CITIZEN FOOD SAFETY

Immediately following the triple crisis, the Japanese government quickly scrambled to respond and create a narrative of the events. Some of the first official reactions came from NISA, Japan's Nuclear Industry and Safety Agency which ordered an initial evacuation area of a 10-kilometer radius around the power plant. The news from other government agencies and formal reaction from the Prime Minister was delayed as news agencies from around the world descended on Japan and began using social media to fill a widening information gap. The lack of information during the crisis, especially for citizens, some of whom were trapped, shows deep problems within the bureaucracy. The desire to manipulate information asserted itself in the chief cabinet secretary's statement that "there was no explosion" while international news agencies were using social media and Google maps to refute this claim.[30] The evacuation zone was then quickly updated on later on March 12 (a full 24 hours after the explosion and meltdown) to 20 km, when the agency informed "the IAEA's Incident and Emergency Centre (IEC) that there has been an explosion at the Unit 1 reactor at the Fukushima Daiichi plant, and that they are assessing the condition of the reactor core."[31] There was still no formal declaration that the reactor had a partial meltdown, which would have required a much larger evacuation and safety agency response. The Kyodo news agency was also reporting that only 6500 were missing and estimated the number of dead at 1700 while a tool created by Google to locate the missing had already logged 60,000 entries. Initially, Prime Minister Naoto Kan said that 100,000 troops would be deployed in the recovery effort, and later in a news conference he told reporters that he believed the disaster was the worst crisis since WWII. This was a rare moment of honesty, that was not repeated, and occurred alongside emerging denials by TEPCO, the corporation responsible for Japan's nuclear power plants which, like many corporations, has a cooperative relationship with government. Unfortunately, during the triple disasters, the government acted on behalf of TEPCO, instead of protecting its citizens and the environment. While the government and industry's response minimized and denied the extent of damage, the citizen reports and activists provided alternative understandings of the crisis. The use of social media, assisted in these efforts and people living in affected areas and those who fled the region, began writing their own narrative online in the form of websites, blogs and facebook pages that documented the human impact.[32]

After the earthquake, tsunami, and nuclear meltdown the government's response could best be described as problematic and at worst, it echoed the response noted by Timothy George regarding the earlier pollution outbreaks when he states that the government lied to its population. Local governments, often lacking the resources to respond to demands made by their citizens had to rely on aid that was slow to arrive from the central government. Oftentimes, the overseas reporting about radiation did not align with information coming from the government, this inconsistency led people living in the Tohoku region to believe that things were likely much worse than they were being told. The grassroots political activity led by mothers in response to the Fukushima Daiichi Nuclear Crisis has been noted by the international press in several prominent outlets and is beginning to be researched by academics as well. A Guardian article highlights activities of the women's movement, noting that "(I)ndeed, groups of women braving a cold winter, have been setting up tents since last week preparing for a new sit-in campaign in front of the Ministry of Economic Affairs."[33] Motivated by their role as caretakers of the home, the women that comprise these movements pressured the government to end the use of nuclear power and evacuate children and families living in areas with high radioactive contamination.

The protest demonstrations against nuclear radiation have been based in Tokyo, utilizing a network of women activists "who have provided the digital framework for organization that has brought together an older generation of anti-nuclear activists, young families, hip urbanites, office workers and union protesters."[34] The use of new media has allowed the women's groups to make connections with mother's groups across the world, garnering support and information gathering for their cause. The information provided by some of these groups has been essential in helping citizens in the Tohoku region understand their reality, in a troubling context of government mis-information and the promotion of industry at the expense of its citizens. The crisis illustrates a pattern noted by environmental scholars about the problematic value given to business by the Japanese government that is oftentimes elevated above the interests of citizens, which is undemocratic.

Safecast is a citizen-science blog providing information on radiation levels across Japan and now the world by distributing its Geiger counter kits. Immediately after the triple crisis, the government downplayed and even lied outright about radiation levels, Safecast was one of a number of groups that emerged to document accurate data, as its website states.[35]

After the devastating earthquake and tsunami which struck eastern Japan on March 11, 2011, and the subsequent meltdown of the Fukushima Daiichi Nuclear Power Plant, accurate and trustworthy radiation information was publicly unavailable. Safecast was formed in response, and quickly began monitoring, collecting, and openly sharing information on environmental radiation and other pollutants, growing quickly in size, scope, and geographical reach. As their website states "Our mission is to provide citizens worldwide with the tools they need to inform themselves by gathering and sharing accurate environmental data in an open and participatory fashion."[36] Using the citizen Geiger counters distributed by Safecast, people living in and around the evacuation zones can monitor radiation levels. The large volunteer organization has recently surpassed 50 million data points and on its website provides a map with real-time data about radiation levels across Japan. Their work has been lauded by National Geographic and empowered Japanese citizens who no longer trust the information coming from government Ministries or TEPCO.

The government response was criticized by online community groups including consumer groups, environmental groups and mother's groups. The Ministry of Health, Labour and Welfare which is responsible for monitoring food through its Department of Food Safety, did establish provisional regulation values for radioactive substances and also established new radiation limits for foods that align with the international Codex standard of 1milliSievert per year (1 mSv/yr). Radioactive cesium (CS 134 and CS 137) have long half-lives and may linger in the environment for many years and both of these radioactive particles have been found in food in Japan. In early March, levels that exceeded MHLW standards for radionuclides in food were found in milk and spinach from the Fukushima area, in Gunma and Ibaraki prefectures, spinach with high Cesium 131 levels were found as well. In June, elevated levels of radiation (CS-134) continued to be found in Shizuoka (green tea leaves) and in fish from Fukushima.[37] Farmers in the Tohoku region had their lives uprooted, government policies were unclear and many people were afraid of buying goods from the area. Although the region is not a high producing area for rice, there are vegetable and dairy farmers. These citizens had to navigate inconsistent and uncertain policies regarding selling their crops. The government stopped shipments of milk and spinach from Fukushima after admitting that levels of radionuclides were high enough to exceed national limits but did not take a more proactive approach and issue a nationwide ban on spinach, which may have eased fears and given people higher confidence in their government.

The single policy change that resulted immediately from the triple disasters with regard to food security was the creation of new guidelines regarding the presence of radionuclides in foods which was adopted in April 2012. This ministerial-level ordinance was created by the Radiation Council of the Ministry for Education, Cultural, Science and Technology (MEXT) which was organized after the disasters. The process through which the council came to the decision to revise its policy was highly politicized. The 1 mSv/yr limit was considered by many to be too high and to do little except to provide relief to citizens that were hoping for government action. Along with revising the standards, the MHLW also revised several other existing laws. The degree to which these new laws protect citizens from the radiation which leaked out of the reactors in Fukushima and penetrated the soil, air, and water in the area is arguable. The government's response to the Fukushima Daiichi crisis, its management of GMOs and relationship with environmental and citizens groups shows a dynamic that fails to elevate citizen politics to formal government. The majority party, the LDP has not shown an interest in incorporating these groups and their interests into its politicking and bureaucrats within the ministries responsible for trade and agriculture use policymaking as a tool for self-preservation instead of representing citizen constituencies that advocate for the environment. The weaknesses of the Japanese government in these areas are not singular, many democracies illustrate similar dynamics but few go as far to manipulate information and co-opt the efforts of citizens.

NOTES

1. Murota, Yasuhiro. 1985. "Culture and Environment in Japan." *Environmental Management* 9 (2).
2. Patrick, Hugh. 1976. *Japanese Industrialization and Its Social Consequences.* Berkeley, CA: University of California Press.
3. George, Timothy, S. 2001. Minamata: Pollution and the Struggle for Democracy in Postwar Japan, 43. Cambridge: Harvard East Asian Monographs.
4. Ibid.
5. Schreurs, Miranda. A. 2002. *Environmental Politics in Japan, Germany, and the United States.* Cambridge, UK: Cambridge University Press.
6. Ibid., 42.
7. Hasegawa, Koichi. 2004. *Constructing Civil Society in Japan: Voices of Environmental Movements.* Melbourne: Trans Pacific Press.
8. George, Timothy S. 2001. *Minamata: Pollution and the Struggle for Democracy in Postwar Japan.* Cambridge: Harvard East Asian Monographs.
9. Broadbent, Jeffrey. 1998. *Environmental Politics in Japan: Networks of Power and Protest.* Cambridge: Cambridge University Press.
10. Schreurs, Miranda. A. 2002. *Environmental Politics in Japan, Germany, and the United States,* 5. Cambridge, UK: Cambridge University Press.
11. Ibid.

12. Ibid., 6.
13. Schreurs, Miranda. A. 2002. *Environmental Politics in Japan, Germany, and the United States.* Cambridge, UK: Cambridge University Press.
14. Mason, Robert, J. 1999. "Whither Japan's Environmental Movement? An Assessment of Problems and Prospects at the National Level". *Pacific Affairs* 72 (2): 187–207.
15. Miller, Alan, S. Reviewer. 1989. "Three Reports on Japan and the Global Environment". *Environment* 31 (6): 25–29.
16. Impoco, Jim. 1992. "Japan's Late Greening: Will the World's Self-Ordained Eco-Cop Keep Trashing Its Own Backyard?" *US News and World Report* 110 (12) (March 16): 61–63.
17. Ibid., 62.
18. Miller, Alan S. Reviewer. 1989. "Three Reports on Japan and the Global Environment." *Environment* 31 (6) : 25–29, 15.
19. Mason, Robert, J. 1999. "Whither Japan's Environmental Movement? An Assessment of Problems and Prospects at the National Level." *Pacific Affairs* 72 (2): 187–207.
20. Ibid.
21. Nakai, Yasutaka. 2001. "Japan's Clean Development Mechanism and the Fight Against Global Warming." *NIRA Review* (Winter): 12–15.
22. Ibid.
23. Lynch, D., and D. Vogel. 2001. *The Regulation of GMOs in Europe and the United States: A Case-Study of Contemporary European Regulatory Politics.* New York: Council on Foreign Relations.
24. Interview. 2017. With Martin Frid, Director, Citizens Union of Japan, *Nishoren.* July 28, 2017.
25. Chan, Jennifer. 2008. *Another Japan is Possible: New Social Movements and Global Citizenship Education.* Redwood City, CA: Stanford University Press.
26. *Japan Agricultural News.* 2016. "Zen-Noh Group and US Producers Band Together for Stable Supplies of Non-GMO Corn". *English Agricultural News* (February 2). Available at http://english.agrinews. co.jp/?p=4216.
27. Robert Putnam. 1998. "Diplomacy and Domestic Politics: The Logic of Two-Level Games", *International Organization* 42 (Summer): 427–460.
28. Biotechnology is Defined as the Application of (a) In Vitro Nucleic Acid Technologies Including Recombinant DNA and Direct Injection of Nucleic Acid into Cells or Organelles or (b) Fusion of Cells Beyond the Taxonomic Family.
29. Interview. 2017. With Martin Frid, Director, Citizens Union of Japan, *Nishoren.* July 28, 2017.

30. Goodman, Roger (eds.). 2002. *Family and Social Policy in Japan: Anthropological Approaches.* Cambridge, UK: Cambridge University Press.
31. Ibid.
32. Freiner, Nicole. 2013. "Mobilizing Mothers: The Fukushima-Daiichi Catastrophe and Environmental Activism in Japan." *ASIA Network Exchange: A Journal for Asian Studies in the Liberal Arts* 21 (1): 27–41. http://doi.org/10.16995/ane.37.
33. Kakuchi, Suvendrini. 2011. "Japanese Mothers Rise Up Against Nuclear Power." *The Guardian* 22 (2) (December). www.guardian.co.uk.
34. Slater, David. H. 2011. "Fukushima Women Against Nuclear Power: Finding a Voice from Tohoku." *The Asia-Pacific Journal: Japan Focus* 9 (1) (November). www.japanfocus.org.
35. Safecast. 2018. "Documenting Radiation Levels." Available at https://blog.safecast.org/. Accessed May 21, 2017.
36. Safecast. 2018. https://blog.safecast.org/. Accessed February 28, 2018.
37. Ministry of Health, Labour and Welfare. 2011. *New Standards for Radionuclides in Foods.* Tokyo: Department of Food Safetey, Pharmceutical & Food Safety Bureau.

CHAPTER 6

Food Sovereignty, Safety and Security: The Role of Rice in Japan and Asia

At the beginning of the current century, in each of the first four years world grain production failed to meet levels of consumption. First in 2000–2001, shortfalls were caused by countries drawing down their stocks impacting 2002–2003 as well. These four years of shortfalls dropped global stocks to the lowest levels in 30 years (Brown 2004).[1] As demographic shifts and evolving trade relationships impact the way states manage their international and domestic politics, Japan confronts key challenges. The supply of food is tightening while global demand and populations rise. It is predicted that the world's population will increase by 3 billion by the year 2050.[2]

The United Nations Food and Agriculture Organization (UNFAO) states that food security exists when all people at all times have access to sufficient safe and nutritious food which meets their dietary needs. Food security is stable then, when people can reach affordable food that meets their caloric needs and is available in a social context that allows them to purchase food without having to cross gender, ethnic, or linguistic barriers that may prevent access. UNFAO uses four dimensions of measures to examine food security, these dimensions are availability, access, utilization, and stability.[3] The current state of food availability globally is insecure, with gaps in the availability of food and problems of malnutrition and growth stunting in Africa, Asia, and Oceania.

While Japan's population is declining sharply those of the East and Southeast nations around them are increasing. China and India are both becoming trade forces as their populations consume more grain and

© The Author(s) 2019
N. L. Freiner, *Rice and Agricultural Policies in Japan*,
https://doi.org/10.1007/978-3-319-91430-5_6

though smaller countries, Indonesia, and the Philippines both have younger populations that are growing. While the competition for food is increasing, the growth of the world's grain supply is slowing for a number of reasons. As Lester Brown (2004),[4] a noted environmental and climate activist and former President of the Earth Policy Institute notes, the ability to produce more grain is unlikely because the backlog of unused agricultural technology is shrinking, available cropland is being converted to nonfarm uses, rising temperatures are shrinking harvests, aquifers are being depleted, and the amount of available water for irrigation is being diverted to cities.

Food security and food sovereignty are now issues that are more clearly on the global agenda that they have been for some time. The vulnerability of the global food supply chain has been highlighted by recent crises caused by dramatically shifting weather patterns as a result of global warming. The UNFAO notes in its most recent Food Security Report[5] that currently 815 million people in the world are hungry and with the expected growth in population that will occur in the next 30 years global food production would need to increase by 50%. This rise in production is unlikely, for reasons that will be discussed in the following paragraphs. World hunger is on the rise now after decades of declines, undernourishment has increased in the last two years and while it is not clear that this is a trend, 11% of the global population now suffers from undernourishment because of conflict which prevents food from reaching needy citizens and oftentimes these conflicts are compounded by climate-related weather changes. In Asia, the uptick in undernourishment is most pressing in Southeast Asia, where the number increased from 9.4 to 11.5% between 2015 and 2016 returning to levels approximating 2011. Southeast Asia contains the largest number of people undernourished in Asia, in countries such as Sri Lanka, Indonesia, Malaysia, and parts of southern India. Japan is a developed country with food security that is currently stable, it is also the largest contributor to UNFAO programs including recent projects that share rice growing technologies with African countries. Countries with stable food security may also be concerned with food sovereignty, although stable food supplies may exist for most developed countries these systems are stable within relatively short time spans. Most markets and groceries in the developed world have food stocks that are enough to sustain only a 48-hour period, beyond this window if transportation problems, weather catastrophes or other events prevent food supplies from reaching the market there is nothing to fall back on.

As Japan moves forward with the CPTPP Trans-Pacific Partnership without the commitment of the United States, a desire to reestablish itself as a power player in the region seems likely. Prime Minister Abe's assertion that Constitutional reform will occur soon and become a part of his leadership legacy means that security relationships in Asia will respond to this change, Asia is likely to become a more difficult region to predict in terms of security as these changes occur. With the stability of the Asian region threatened the ability of countries there to maintain stable food supplies and establish food sovereignty becomes more difficult.

THE GLOBAL TRADE IN FOOD

The story of food security is intimately related to the way in which agricultural products and food are traded. The supply of global food and trade in agricultural products is important in understanding how and why food security has become a relevant issue. The story is dramatic and underlying it are powerful dynamics in the role of states in constructing and adapting to a global food trade that is dominated by large multinational firms based mostly in Europe and North America. The trade in agricultural food systems is relatively recent, undergoing a substantial increase since the early 1990s when the volume of global trade expanded faster than the volume of global production. Trade volumes increased 49% between 1990 and 2001, production increased only 25% and with this shortage in production to meet global demand came a restructuring of food systems along political lines with states supporting the opportunity for profit.

Asian food imports were worth 60 billion dollars by 1990 and made up 32.1% of the worlds food trade.[6] By 2000 the number jumped to 98 billion and 35.8% of the global total. It became possible to discuss an East Asian food import complex that was provisioned by northern agriculture or low-cost production sites in the Asia Pacific. Currently, the trade in global agricultural products is worth $600 billion annually with a large portion of this trade occurring in Asia.[7] A large portion of agricultural trade has been driven by Japan, the largest and most affluent consumer market for food imports, whose purchases of food increased 300% from 1980 to 2000 (from 149 billion to 48.6 billion). It is the largest net importer of food in the world, making up nearly 20% of the total in the global food trade and 50% of food imports just in Asia. Japanese self-sufficiency of its citizens' caloric intake during this period dropped drastically from 79% self-sufficiency in 1960 to only 41% in 1997.[8] Japan's imports of dairy (cheese) account for a large portion of these imports as more than half of all imported dairy products are some form of cheese.

A number of researchers note that the lack of a rules-based system of trade governing the global food industry has been the dream of many, however, while both the Uruguay and Doha rounds of negotiations attempted to tackle large portions of the trade in agricultural products the achievements of the Uruguay Round were modest. The Doha negotiations collapsed under the number of issues that were on the agenda and the national interests of many countries, especially in the developing world who wish to maintain protections for farmers.

The issue of food security is becoming something that is taken more seriously by countries in the wake of global droughts and drastic weather changes that affect crop prices. The world's reliance on grain is likely to grow in the coming decades, both climatologists and agricultural experts predict that global grain shortages are likely to continue and that food security will become increasingly important. Today, Japan imports 40% of its citizen's diets, making it one of the least secure nations in terms of food. Currently, the world's largest grain exporters (Canada, the USA, Australia, and Brazil) supply Asia with most of its rice, wheat, and barley. In 2011, when a drought in China affected Chinese grain, it had to go onto the global market and that year China became the global grain market. This year, North Korea is experiencing its worst drought in 16 years and will have to rely on imports from China.

THE GLOBAL OUTLOOK FOR FOOD SECURITY

Food Security is a relatively new concept in international policymaking literature, the term was first used in the mid-1970s during the global food crisis. The concept at this time was most focused on the supply and stability in price of food, reflecting the context of the food shortages, and steep prices associated with the changing organization of the global food economy. After the mid-70s food supplies stabilized but the concept was further refined by organizations such as the UNFAO and the World Bank (its 1986 report, Poverty and Hunger brought in concerns regarding chronic food insecurity related to poverty and transitory food insecurity caused disaster, economic collapse, and/or conflict). The UNFAO currently defines food security as follows:

(F)ood security exists when all people, at all times, have physical, social and economic access to sufficient, safe and nutritious food which meets their dietary needs and food preferences for an active and healthy life. Household food security is the application of this concept to the family level, with individuals within households as the focus of concern.[9]

As noted in earlier chapters, the Japanese public is concerned with food safety, food soveriegnty, and food security issues, stemming from the 1918 rice riots and protests that followed, to modern day with concerns about food safety that emanated during the heavy phase of industrialization and large pollution outbreaks which contaminated fish, water, and other products in Japan's food chain. Recently, the triple disasters once again highlighted the Japanese government's lack of infrastructure regarding the safety of its food supply. During the Fukushima-Daiichi crisis, what followed was a movement among mothers concerned about the radioactivity present in the food their children were eating. The triple disasters highlighted the lack of appropriate policies governing the safe supply, distribution, and transportation of food as well as the non-existence of government agencies to oversee food safety. The response to the crisis which was discussed in Chapter 5 also shows the vulnerability of Japan's history of reliance on food imports in order to meet its food demands.

Japan was the first country to achieve a take-off in agricultural production, this was a result of the green revolution which applied science to both grain genetics and agronomics. During the 1880s, Japan developed the dwarf variety of rice which allowed it to drastically increase its production of rice. The genetic contributions to developing the dwarf rice plant were changes made to increase the share of the plant's photosynthetic product going into the seed, basically shifting photosynthate from the leaves, stem, and roots to the seed to maximize yield. Shifting photosynthate to seed makes the crop more efficient. Longer stalks are necessary for plants that are competing with other plants for sunlight, creating a shorter stalk makes plants that are grown in large beds better able to use their exposure to sunlight and convert it to photosynthate. Farmers that breed plants call this the harvest index. The key to increasing the harvest index in Japan was the use of genetics, agronomic improvements, and synergies between the two[10] (Brown 2004) to introduce the dwarf gene into rice. Reducing the stem length, reduced the share of photosynthate going into the straw, the gain in yield was equivalent to the loss in straw weight. Over the years, since the green revolution, plant breeders have increased the harvest index (share of photosynthate going to the seed) but have not been able to improve the efficiency of photosynthesis. The total photosynthate of today's plants has essentially remained unchanged from their wild ancestors, what has changed is the harvest index or the amount of photosynthate going to the seed.

Along with genetic improvements, Japan has used changes in cropland management, irrigation, fertilizer, and controlling pests, diseases, and weeds. Japan uses wet rice agriculture to deal with weeds, by flooding rice fields early in the plant's growth, competition from weeds is prevented. Water management is central to rice agriculture and to some degree, this trait has been a crucial part of its domestication. Using wet rice agriculture to control weeds instead of chemicals is a low technology solution from which all farmers can benefit. The geography and climate of Japan make it a perfect location in which to grow rice, as many authors have noted.[11] The distribution of water in small fields ensured the productivity of the crop, although the small size of Japan's rice paddies has been criticized for preventing more productive large-scale agriculture. However, the maintenance of small paddy growing has maintained other features of Japanese rice culture, including ownership of small plots by families and the preservation of the water management system. The communal sharing of water is a distinct feature of Japanese wet rice growing, the intensive networks of irrigation canals, and red diversion devices mark the country landscape (see Figs. 6.1 and 6.2).

The water canals are supplied by a variety of different water sources, including rivers, mountain streams, and holding ponds dug out and fed by springs. These devices are managed by local irrigation cooperatives that function based on traditional water control procedures rather than formal law in many cases. There are both formal and informal cooperative arrangements that allow beneficiaries of the cooperative to take their allotment of water when necessary.

During the Meiji era, water became national property, and since 1896 the allocation of water was administered by the state while the agricultural use of water has been managed by local land improvement districts. This unique feature of Japanese wet rice agriculture has been noted by Eyre[12] the author of the only study on these irrigation systems to be published in English. Eyre's study focuses on a single water-sharing district in Okayama prefecture, where he studied the local cooperatives

Fig. 6.1 Irrigation system, Toyama prefecture (Photo by author)

Fig. 6.2   Irrigation Canal, Toyama prefecture (Photo by author)

which manage the many diverse water sources and diversion devices which manage the water systems. He estimated the number of these local cooperatives at well over 100,000 in the 1950s.[13] The members of these districts have agreements with the district as beneficiaries that pay water charges. These water systems maintain a traditional system of water control that relies on an allotment according to established procedures and orders of cooperation. More recently, Smil and Kobayashi noted that by the year 2000 there are more than 1000 dams and 42,000 water channels that have the potential to deliver more than 60 billion cubic meters per year of water, two-thirds of the country's total water usage.[14]

These irrigation systems are still in existence and are operational, the pictures above are taken from water management devices in Toyama prefecture, where local water management cooperatives oversee most of the devices while many others are managed by local informal groups. It is likely that some of these irrigation devices are either not working regularly or are defunct as increasing numbers of Japan's paddy fields lay fallow. The water management devices then are not only part of Japan's cultural and performative rice agriculture, it is also a part of the history to which it is linked in Japan. As efforts are made to increase the size of Japan's rice fields, a policy goal of the MAFF outlined in several of its recent publications, these water management systems will have to be destroyed to make way for the change.

Along with managing water for weed control, the productivity of dwarf rice is based on a precise spacing of plants in carefully tended rows. Dwarf rice is just one plant variety that was included in the green revolution, which began in the United States with researchers such as Dr. Wallace and Dr. Borlaug who studied ways to make plants more productive and resistant to disease with great success. The green revolution seed technologies were advocated by the World Bank and International Monetary Fund, the international institutions funding development, in their aid packages to countries in the developing world. The gains in productivity in Japan were also made in Europe and the United States. Between 1950 and 1976, the years during which green revolution seed technologies and agronomic methods were utilized by the developed world increased the world grain harvest twofold.

In Asia, because of the history with rice shortages and food scarcity, and in Japan noted in Introduction, food security and food sovereignty are a concern for the government. The most important factor when discussing food sovereignty in Asia is rice. For the dominant

Asian economies of India, China, Thailand, Indonesia, Japan, North Korea, South Korea, and the Philippines, rice is synonymous with food, and access to stable supplies of rice is equivalent to food sovereignty. There are other factors however that complicate this story. These factors include rising incomes in the developing world and in Asia, especially in China and India, that have created new dietary preferences and an increasing demand for meat proteins. These dietary shifts are having an impact on the global supply of grain and meat. Also, land and water resources are finite, increasing prices for fertilizer and fuel for transportation and storage are additional factors that add to food security concerns for countries in Asia. Moreover, the incidence of climate crises is rising along with global temperatures increasing the likelihood of droughts, floods, hurricanes, and tsunami. These drastic weather patterns and the crises they bring are concerns for which most countries are unprepared.

During the most recent food crisis in 2007 and 2008, the food price index rose by 54%, from January 2007 to June 2008, for consumers the real price of rice tripled. In Asia, poverty and food insecurity are trends that coincide with economic changes such as rising incomes, the income gap is widening in many countries, including in India and China the two Asian giants. Writers and researchers discussing food security note the "two faces of Asia" because although Asia's share in global consumption is rising, per capita consumption is still below average. Patterns of food consumption and production are changing and along with the drive for food sustainability, Asia's share of global consumption will dramatically alter the global food system in the coming decades. The People's Republic of China (PRC) and India account for nearly 40% of the global population and India's population will surpass China's in this decade. Its growth will alter food systems in the region and the world.

The change in the trade of agricultural products and emergence of a trade in food which could truly be called global began in the 1980s after the Uruguay Round of negotiations regarding the General Agreement on Tariffs and Trade (GATT). During 1988 and 1989, Japan and the United States met for a round of negotiations that produced concessions by the Japanese to allow imports of beef and citrus from the United States. These concessions prepared the Japanese for a gradual set of reductions in agricultural protection which increasingly opened its market to food imports. In 1995, accession to the World Trade Organization

(WTO) meant changes in Japan's legal framework for supporting agriculture and a review of its Agricultural Basic Law of 1961, resulting in its adoption of the New Policies of 1998 and 1999 which restructured its support of agriculture in order to meet up with its commitments to the new WTO. Likewise, other countries did the same and around the world, agricultural protections underwent major changes.

In the 1990s, agricultural policy reform and liberalization were discussed openly in Japan in the context of a set of broad political debates caused by the burst of Japan's "bubble economy" and the ensuing economic crises. The crisis precipitated questions about the value of Japan's commitments to its farmers and the cost of the support that was provided, especially to rice and dairy farmers. The debate about the merit of these policies in the midst of economic crises was amplified by challenges from the urban population whose political voice was underrepresented because of Japan's electoral systems favoring of rural areas (a source of the traditional base of the Liberal Democratic Party (LDP)) discussed in Chapter 4. The debates about agricultural supports and the LDPs uneven representation of rural areas was balanced by concerns about the country's food self-sufficiency and decreasing ability to source its demands by domestic producers. The trade ministry, MITI also contested farmer protections are argued for the support of modern methods of farming and the support of manufacturing (its constituency). The tensions exposed by these debates illustrate the country's incomplete transition in agro-food provisioning from small family farms to a modernized food industry. The lack of a transition is typified in the agricultural areas that have sustained government support since the WWII era, the dairy and rice sectors. As the focus of this book is rice, its role in self-sufficiency will be the focus of the following sections of the chapter. Moreover, the debate also shows the continuing tension between the MAFF's support of rice growing as a cultural and performativity activity which is connected to the political interests of LDP norinzoku politicians and bureaucrats at the expense of consumers and farmers. Consumers would benefit from increased competition in Japan's food market and farmers would benefit from policy changes that would promote more competitive, efficient farming. Neither of these changes are likely given the entrenched interests of policymakers who continue to prioritize political interests over those they have a democratic obligation to represent.

FOOD SAFETY AFTER THE TRIPLE DISASTERS

Immediately following the Fukushima-Daiichi Crisis, issues with food safety and security became prominent immediately following the detection of radioactivity in plants in the Fukushima area. These levels were higher than those cited in the National Safety Commission of Japan's (NCSJ) guidelines of 1998 which had been established earlier. The Ministry of Health, Labor and Welfare was responsible for determining safe levels for food according to its charge in the Food Safety Basic Act. The MHLW is tasked with consulting an independent Cabinet Committee, the Food Safety Commission in advance of setting any limit for food. However, the quick unfolding of food safety concerns prevented the ministry from taking this step and instead it issued numbers itself, adopting the older guidelines as "provisional" values. The provisional values set the annual radiocesium dose at 5 mSV (millisieverts) with a limit of 200 Bq/kg for milk or water and 500 Bq for all others. The annual dose for radioiodine levels was set at 50 mSV with 300 Bq for milk and water and 2000 Bq for all others. On March 21st, the MHLW ordered that all agricultural products in prefectures in the contaminated areas would be banned from shipment or sale. This restriction was applied to entire prefectures, and punished some farmers far from the contaminated areas, existing food labeling laws required prefectural level origin labeling, making it difficult to control food at the subprefectural level. On April 4th, the MHLW allowed a subprefectural designation for the shipping restrictions and a method for lifting the ban after a banned product passed testing for three weeks which helped to alleviate some of the restrictions although consumers still avoided products from the Fukushima area.

If any food exceeded these values, it was ordered that the entire lot would be destroyed as a violation of the Food Sanitation Act. Later on, when the Ministry finally consulted the Food Safety Comission, the commission faced an important dilemma, would it shift the numbers upward or maintain what the MHLW had already established? Indicating that a higher level was appropriate would open the MHLW to criticism from the public, during an already highly contentious period following the crises. However, food control in Japan warranted that the levels set by the MHLW constituted a cancer risk (albeit one that was of low probability) deemed unacceptable by these standards. These numbers were also in conflict with the limits set by the International Commission on

Radiological Protection, which issued a report dedicated to the people of Japan on April 4 and recommended that a limit set at 1 mSV/yr was consistent with past experience of radiological disasters and mitigation of long-term effects.

The government screening program was the only basis for consumer choice, but it offered few guidelines or assistance to standardize the directives it gave to prefectures. Therefore, at the prefectural level, authorities were left to improvise so screening was implemented very unevenly. In June, the government finally issued a timeline to sample food but there are only a few dozen officially certified labs for radiological inspection and the number of food that had to be screened meant for very slow processing and backlogs of food requiring screening. In the months immediately following March 2011, a number of major violations affected public confidence. For example, beef contaminated with radiocesium was sold without screening. When listing agricultural products to be prohibited for use, government authorities had not included rice straw used for fodder, assuming that it had been stored inside when in practice many farmers leave straw out on fields to dry exposing it to radioactive contamination. When it was found to exceed limits, the "radioactive beef" was removed from stores and destroyed but the impact had already affected the public. The system to control food was over centralized, relying too heavily on numbers which themselves were unreliable given the debatable scientific evidence and lack of consensus on risk.

Conflicting reports about radiation levels and the government's seemingly wavering response, led many mothers to worry about the health of their children following the disasters, especially health effects of radiation especially radiation levels of the food offered in public school lunches. Some mothers began making lunch for their children leading to a situation where their children felt that had to conceal the food they were eating because their mothers considered what their friends were eating to be "potentially impure and unsafe."[15] Many mothers began to organize themselves and formed groups to share information about radiation levels as well as their experiences with finding safe food sources. Food within the unsafe zone was considered irradiated and could not be sold even though some local farmers did collect and eat the food themselves. In a poignant story in the first volume of research to come out after the crisis Gill[16] describes his visit to the local headman in Nagadoro (a hamlet in the village of Iitate, in the evacuation zone) who presented him with a gift of mushrooms:

He could not sell them, because they had been deemed dangerous. But he would eat them himself, and give them to other people to eat. Refusing to accept the mushrooms would be hurtful to him—perhaps implying that I thought his village was defiled. In contrast, accepting the mushrooms would be a gesture of solidarity with Nagadoro. So I took them back to my home in Yokohama. I intended to each them, but that night the TV news said that radiation was 14 times higher than the government approved level had been found in shiitake mushrooms from Iitate. I carried my five kilos to a nearby patch of wasteland, and sorrowfully cast them into the long grass. It was a year later that I finally found the courage to admit to Shoji-san that I had thrown them away. I think he forgave me.

Another writer in this same volume of essays, Ikeda Yoko,[17] notes the personal balancing act that people performed every time that they made a choice about what to eat and what not to eat. People felt at odds with the inconsistent and at times, unbelievable information that was foisted upon them, alongside the endless running choices they had to make. Some schools and villages began to provide citizens with Geiger counters to measure radiation levels and eventually a site set up by the Fukushima prefectural government, allowed people to look up levels of radioactive materials in food. In the months following the disasters, contamination of food was reported in products all over Japan although the methods used to calculate and interpret these radiation numbers were disputed. The avoidance of food from the Tohoku region by many Japanese has become more focused on those goods coming from the Fukushima area specifically but even within Fukushima radiation levels vary widely across villages and across products.

Even well after the triple disasters, radiation continued to be found in food tested in Fukushima and Ibaraki. In June of 2011 for example, Cesium 134 was found in tea leaves and also in fish (fat greenling, brown hakeling, and ayu/sweetfish) from Fukushima. Testing of foods can be somewhat misleading because some foods are more susceptible to radiation than others, milk and fish, for example, often exhibit high levels of radiation. The achievements of the movements have been notable, the Japanese government amended a number of laws to address the issue of nuclear radiation found in food, including setting new limits on radionuclides in food, revising the Ordinance on Milk and Milk Products Concerning Composition Standards, Revising the Notification on Designation of Radioactive Substances, and revising the Notification on Partial Revision of the Specification and Standards for Food, Food Additives.

Food Security and Sovereignty in Asia and Japan

Rice plays a special role in the Asian diet, it is the primary staple food and for Japan it is at the center of its traditional *washoku* diet. Rice is also the staple for China and South Korea, countries with rising populations. The price increases on rice during the food crisis in 2007 and 2008 were alarming for countries across Asia, and especially Japan whose history of protests regarding access to rice are part of Japan's agricultural policy machinery. As the Asian Development Bank notes the price of rice has been key in the industrialization of Asia as a staple for wage earners,

> Its economic and political importance is indisputable, reflected largely in heavy government market intervention and proclaimed self-sufficiency across Asia's rice consuming countries. But with the global rice market relatively thin compared with other crops, rice price volatility is far more pronounced than for most other staple foods.[18]

The importance of rice in Asia cannot be underestimated especially for those nations that rely on it as a food for its poor and low-wage earners. In terms of countries importing rice in Asia, Japan is considered a wealthy importer. Japan's average income is higher than many of its neighbors and though currently it is experiencing an economic recovery as a result of the Fukushima-Daiichi catastrophe, it has already experienced rapid industrialization and extreme economic growth making it the poster child for successful development in the region.

The rice market has a number of characteristics that make it especially volatile and opaque when it comes to trading this commodity. This volatility is largely a result of two factors. The first is interventions by governments driven by political decisions about their imports and exports, the second is the private nature of transactions that are not publicly reported, making calculations difficult. The structure of rice production is also a factor, rice as yet is not farmed by large agro-corporations, it is farmed by small producers, and traded and processed by smaller businesses that occur with other grains (e.g., corn, which is farmed and traded by a small number of large agribusinesses and traded and processed by corporations that report their transactions publicly). Moreover, rice can be stored for nearly two years without deteriorating. The storage of rice by countries like Japan constrains the rice market, making political calculations difficult altering availability substantially. Finally, there is no deeply traded futures market for rice, which can provide some hedging that can cushion potential shortages. Futures trading began in other agricultural commodities and as Silber[19] (1985) argue they provided important economic benefits as futures trading helped to standardize future delivery and created a space for disseminating price information. The absence of a futures market for rice adds to its volatility. These factors create a situation where political decisions are made in an information vacuum. There is nearly no data on price expectations or their market consequence, the rice market operates with very incomplete information about supply and demand making it one of the most unstable, subject to miscalculation.

The structure of the rice market also adds to its unpredictability, the market is thin and concentrated. Thin markets are those that have a small number of buyers and sellers making few transactions, these types of markets are more risky making prices difficult to predict and therefore apt to be unstable, with drastic and sudden price jumps and drops. The thin-ness of the market means that only 7–8% of the rice that is grown is sold, small quantity shifts in these low numbers can greatly affect prices, if traders don't have the resources and savings to sustain the price shocks, prices on the market can be driven by panic. The rice market is also concentrated because it is limited to a small number of countries that purchase and sell rice. Although it is exported by many nations only five countries are the top exporters, accounting for 80% of the rice market— Thailand, Vietnam, the USA, India, and Pakistan. If a single one of these countries alters how much rice they sell for export, it will have a large impact on rice prices. The biggest purchasers are the Philippines, Japan, China, India, Vietnam, South Korea, and Indonesia.

During the global food crisis of 2007–2008, both Thailand and India took measures to restrict their exports of rice after India's drought. The drought led the Indian government to take measures to ensure that that they had enough rice for its public distribution system which offers food to poor people at subsidized prices. The drought led to a shortfall in wheat, the government tried to fill the gap by buying more rice while at the same time other countries were also switching from wheat to rice.[20] This caused the price of rice in India to rise sharply, prompting the government to take action and ban its export. Thailand responded to the drought based on its own concern about feeding its poor. Others in the region panicked and the prices went soaring. The panic made governments wary of their lack of food security and dependence on imported rice, leading some to stockpile rice and commit to self-sufficiency. The panic was not based on the real market price of rice, but on the emotional response to government announcements. When Japan announced that it would release some of its stored rice to be traded on the market, prices dropped well in advance of the rice actually reaching the marketplace. As McMahon states "(A) series of government decisions, although perhaps rational in their own terms, led to collective panic."[21] This vignette illustrates the impact of government policies on the rice market and potential for rice prices and limited access to rice as demographic changes occur in Asia. The growth of other Asian nations, especially those with younger populations, and the development of less wealthy rice importers will impact Japan's access to rice and other foods as well as their price (Chart 6.1).

Of the top nations that consume rice, Japan is a relatively smaller and richer importer as noted above, but recently its share has been rising. The chart above illustrates that rice consumption per capita has remained relatively stable within the last decade, with small increases and even some decreases. Japan's share of rice imports compared to China is small, but China's presence on the market affects Japan. As Japan's population ages and there are increasing budgetary constraints on its government and its population, rice will continue to be an important staple food. As its own MAFF has noted, global demand for food will increase, especially in Asia, the projection for global food demand is a rise of 60% from 4.5 billion tons in 2000 to 6.9 billion tons in 2050[22] (United Nations 2012).

Chart 6.2 illustrates the story more dramatically when comparing the import quantities across the largest importing and exporting countries, the volatility in the rice market is more clearly shown. In 2009, China was still importing large quantities of rice after the Indian drought, drop their imports fall off sharply as the market stabilizes. The Philippines and Indonesia are the biggest import customers of rice, their movement on the market heavily influences prices, as does India's. Highly fluctuating moves on the rice commodity market spell trouble for prices and make it difficult for countries to navigate policy in order to accommodate shifts. After the global food crisis in 2007–2008, many countries created initiatives to address their dependence on imports, including Japan.

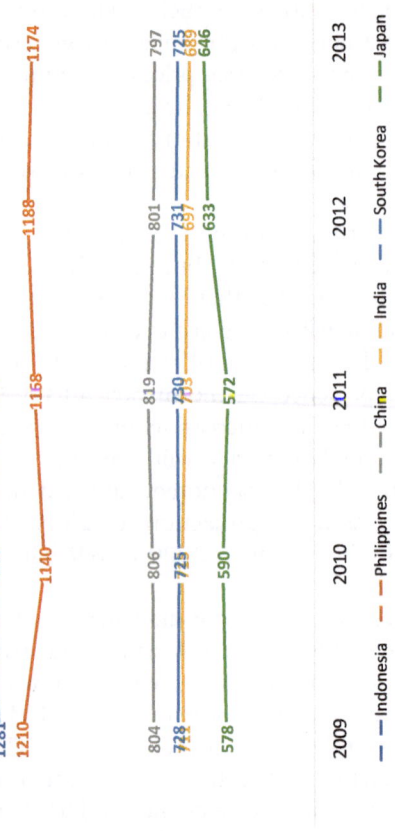

Chart 6.1 Rice consumption in Asia

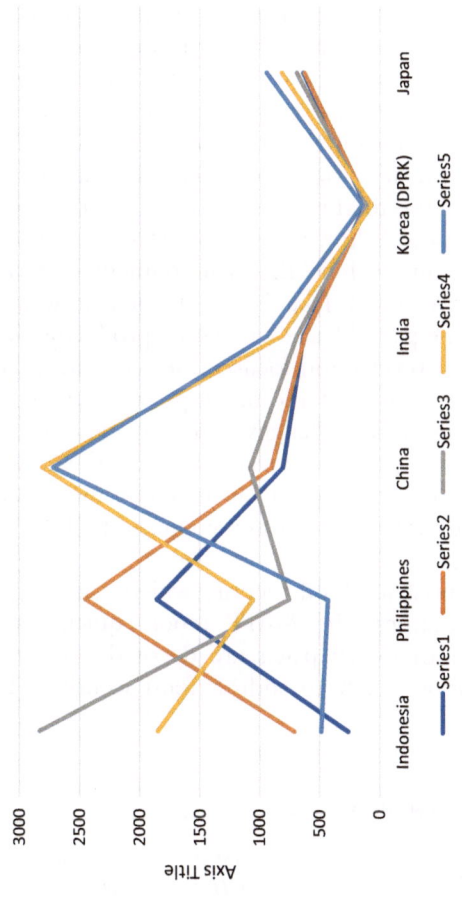

Chart 6.2 Import quantity of rice (1000 tons)

FOOD SECURITY AND POLICY IN JAPAN

Currently, Japan's food sovereignty is low, it imports about 40% of its food supply and the MAFF has set a goal of raising food self-sufficiency to 45% on a calorie basis and to 73% based on production values by 2025. A number of policies are being promoted by the MAFF to bring Japan to higher food self-sufficiency and some argue to maintain the MAFF's ability to intervene in Japan's agricultural economy (Table 6.1).

The MAFF is addressing food security and food sovereignty issues through its programs and has even calculated its own "Food Self-Sufficiency Potential" a measure that it created to determine the country's ability to provide its citizens with enough calories by domestic production alone. As the MAFF notes in one of its own pamphlets, the food self-sufficiency ratio is low, but so is its food self-sufficiency potential. The causes for this include the abandonment of farms and large areas of unused farmland in Japan's countryside as well as trends in Japan's diet. The same pamphlet notes that by producing sweet potatoes instead of the current production focused on rice, wheat, and soybeans, Japan could produce its own domestic crops to meet its caloric needs. This would necessitate drastic changes in the Japanese diet, however. As the chart below shows, potential meals based on this new food secure diet vary greatly from the diet most Japanese consume today, and from the traditional washoku diet which was recognized by UNESCO and discussed in Chapter 5.

Despite the potential for food security which can be guaranteed by producing sweet potatoes, the MAFF report quickly abandons this notion and the rest of its 19 pages are devoted instead to outlining its three pillars of food policy to address food security: (1) Expanding

Table 6.1 Japan's food security

Year	Energy supply adequacy	Protein supply	Cereal import depend	Food supply variability
2007–2009	113	87	76.8	21
2008–2010	111	87	77.3	27
2009–2011	111	86	77.1	44
2010–2012	112	87	76.9	43
2011–2013	112	87	75.8	43

Exports; (2) Securing Stockpiles; and (3) Increasing Domestic Production. The details are a bit fuzzy on how the country will realize total food security through these policies given what is noted on the first page but a number of initiatives are highlighted. The promotion of local and regional products across Japan which includes a new mark that shows products are local is intended to increase the domestic purchase and demand for these Japanese products.

Moreover the promotion of the Japanese diet to the rest of the world is also noted for its potential to increase domestic production. The restructing of farmland and consolidation of farms with regard to rice is a key feature of the report and was also noted in my interview with a member of the MAFF staff. Consolidation of farmland would end small paddy production and will arguably increase the efficiency of rice production. Farmland consolidation has long been a goal of the MAFF and part of the policy talk among Japanese agricultural policy wonks but across many decades it has never been realized. Another notable aspect of the Ministry's efforts related to food security includes the use of high technology methods such as robotics and hydroponic growing, which reduces the carbon footprint.

Security Stockpiles is another topic also included in the MAFF pamphlet as one method of assisting in food security which is never explained or discussed beyond the brief mention that it receives when being listed as one of MAFF's "three pillars." Japan's storage of rice is a decade old policy that will continue and perhaps even expand. In every prefecture across Japan, there are many rice storehouses that are also called "country elevators." The one that I visited in Toyama prefecture, though empty when I visited it had a massive potential that is only partially utilized, see further discussion and a photo of this elevator in Chapter 4.

Concerns about food security have prompted activity on the part of Japan's environmental and consumer's movement that has criticized the MAFF's response as reactive rather than proactive and one that is motivated by corporate interests rather than those of consumers. For example, this spring the diet abolished a law that controlled the seeds used by farmers as well as supplementing its cost. The law (which was also discussed in Chapter 3) was the legal foundation for Japan's agricultural experiment stations, which create budget requests for prefectural governments' seed expenses. The experiment stations designate seeds that are then recommended to local farmers and initiate the budget requests which assist with production costs of the seeds that are sold to farmers at a low cost. The seeds that are regulated by the experiment stations have all been grown domestically but the abolishment of the law means that private companies may produce and sell seeds that come from outside of Japan. Moreover, a Japan Times editorial argues that seeds are likely to become more experimental as the abolishment of the Seed Law undermines the budget requests that experiment stations create on behalf of prefectural governments.

The abolition of the Seed Law paves the way for the entry of large multinational agricultural corporations to produce and sell seeds to Japanese farmers thereby potentially undercutting domestic seed production. It is likely that if Japan follows this path, Japanese farmers may experience the problems documented widely by other farmers in the developing world in countries like India. Once multinational businesses began selling their seeds there, farmers were forced to rely on them more and more because private sector seed technologies are nonproducing (they are largely filial one, F1 hybrids) which means farmers must purchase seeds every year instead of collecting next year's seeds from this year's planting.

The potential for corporate inroads into Japan's seed market is problematic for many consumers and farmers who are worried about the protection of the genetic resources and technology of Japan's domestic seed production. Their concerns are justified, given the history and criticism of agribusiness that have exploited their ability to develop and market seeds to local farmers and growers solely for the purpose of profit. Japan's heavily regulated food supply may not be ready for this opening to compete from foreign business. Abolition of the Seed Law likely means that Japan is giving up control and protection of the seeds that its farmers grow to corporate control and this may have a long-term negative impact on Japan's food security.

The MAFF has taken on the issue of food sovereignty as one of its missions, but it must also take serious Japan's commitment to international agreements which increasingly advocate for the opening of countries domestic to corporate interests. Moreover, policies on food sovereignty issues especially when related to rice are often made reactively, rather than proactively. The current government interventions work to make the rice market difficult for global traders to predict which adds to its instability and the likelihood of this changing for the better are unlikely. An additional factor to add to the complexity in Asia are the increasingly important roles of both China and India, and how these two nation states will manage their food security and sovereignty over the next decade. The actions that China and India take and the evolving trade negotiations between Japan and the United States that will commence in Fall of 2018 or Spring of 2019 will be watched closely by scholars of the global food trade. There is much as stake as these agreements evolve and they have the ability to affect not only the long-term norms of trade in the region but also the region's food stability including stable access to rice.

Notes

1. Brown, Lester R. 2004. *Outgrowing the Earth: The Food Security Challenge in an Age of Falling Water Tables and Rising Temperatures.* New York, NY: W. W. Norton.
2. Ibid.
3. FAO, IFAD, UNICEF, WFP, and WHO. 2017. *The State of Food Security and Nutrition in the World 2017: Building Resilience for Peace and Food Security.* Rome: FAO.
4. Brown, Lester R. 2004. *Outgrowing the Earth: The Food Security Challenge in an Age of Falling Water Tables and Rising Temperatures.* New York, NY: W. W. Norton.
5. FAO, IFAD, UNICEF, WFP, and WHO. 2017. *The State of Food Security and Nutrition in the World 2017: Building Resilience for Peace and Food Security.* Rome: FAO.
6. Ibid.
7. Josling, Tim, David Orden, and Donna Roberts. 2010. National Food Regulations and the WTO Sanitary and Phytosanitary Agreement: Implications for Trade-Related Measures to Promote Healthy Diets. In *Trade, Food Diet and Health: Perspectives and Policy Options,* ed. Corinna Hawkes, Chantal Blouin, Spencer Henson, Nick Drager, and Laurette Dube. Oxford: Wiley-Blackwell.
8. Ibid.
9. FAO, IFAD, UNICEF, WFP, and WHO. 2017. *The State of Food Security and Nutrition in the World 2017: Building Resilience for Peace and Food Security.* Rome: FAO.
10. Brown, Lester R. 2004. *Outgrowing the Earth: The Food Security Challenge in an Age of Falling Water Tables and Rising Temperatures.* New York, NY: W. W. Norton.

11. Kumakura, Isao (ed.). *Washoku: Traditional Dietary Cultures of the Japanese.* Tokyo: Ministry of Agriculture, Fisheries and Forestry (MAFF). Available at http://www.maff.go.jp/e/japan_food/washoku/. Accessed December 3, 2017.
12. Eyre, John D. 1955. "Water Controls in a Japanese Irrigation System." *American Geographical Society* 45 (2) April: 197–216.
13. Ibid.
14. Vaclav, Smil, and Kazuhiko Kobayashi. 2012. *Environmental Impacts: Land, Water, Nitrogen.* Cambridge, MA: MIT Press.
15. Morioka, Rika. 2013. "Mother Courage: Women as Activists between a Passive Populace and a Paralyzed Government." In *Japan Copes with Calamity: Ethnographies of the Earthquake, Tsunami and Nuclear Disasters of March 2011,* ed. Tom Gill, Brigitte Steger, and David H. Slater. Bern, Switzerland: Peter Lang Press.
16. Gill, Tom. 2013. "This Spoiled Soil: Place, People and Community in an Irradiated Village in Fukushima Prefecture." In *Japan Copes with Calamity: Ethnographies of the Earthquake, Tsunami and Nuclear Disasters of March 2011,* ed. Tom Gill, Brigitte Steger, and David H. Slater, 235. Bern, Switzerland: Peter Lang Press.
17. Ikeda, Yoko. 2013. "The Construction of Risk and the Resilience of Fukushima in the Aftermath of the Nuclear Power Plant Accident." In *Japan Copes with Calamity: Ethnographies of the Earthquake, Tsunami and Nuclear Disasters of March 2011,* ed. Tom Gill, Brigitte Steger, and David H. Slater. Bern, Switzerland: Peter Lang Press.
18. Asian Development Bank. 2013. *Food Security in Asia and the Pacific,* 2. Mandaluyong, Philippines: Asian Development Bank.
19. Silber, William L. 1985. *"The Economic Role of Financial Futures."* Washington, DC: American Enterprise Institute. http://farmdoc.illinois.edu/irwin/archive/books/Futures-Economic/Futures-Economic_chapter2.pdf. Accessed November 22, 2017.
20. McMahon, Paul. 2014. *Feeding Frenzy: Land Grabs, Price Spikes and the World Food Crisis.* Vancouver: Greystone Books.
21. Ibid.
22. FAO, IFAD, UNICEF, WFP, and WHO. 2017. *The State of Food Security and Nutrition in the World 2017: Building Resilience for Peace and Food Security.* Rome: FAO.

Conclusions and Future Recommendations: The Loss of Japan's Rice Growing Culture

At the outset of this project, I set out to tell the story of a lifestyle that I saw disappearing on each subsequent visit to Japan. The loss of this lifestyle was poignant for me as someone who witnessed the abandonment of farm after farm in the rural area in which I grew up. While I believe that Japan will also suffer this loss and see the advent of large-scale corporate controlled agriculture, including the farming of rice, I don't see this loss as a dramatic one that is likely to be commented on by the Japanese media or taken up as an issue by rice farmers because it will simply become reality slowly as time progresses.

Japan's policymaking machinery that deals with agriculture is foremost directed by the Ministry of Agriculture, Fisheries and Forestry (MAFF), but the MAFF itself must deal with competition from other ministries and if Japan does not show some recovery from its economic slump soon, their demands to reign in the budget including diminishing agriculture spending is likely. The most recent budget requests by the MAFF were large, but the request was not guaranteed by government which cut some of its spending, resulting in an overall budget decrease of between 20 and 30%. While minor, decreases like these over time can be detrimental to the maintenance of MAFF policies like the Farmland Banks and Core Farmer initiatives discussed in Chapter 3. In order to continue

© The Editor(s) (if applicable) and The Author(s) 2019
N. L. Freiner, *Rice and Agricultural Policies in Japan*,
https://doi.org/10.1007/978-3-319-91430-5

to guarantee its own existence, the MAFF will need to keep requesting large amounts from the government and will need to justify the legitimacy of those requests. After detailing the dynamics in this book, it is my belief that the MAFF's ability to do this in the future may be in jeopardy.

In the past the MAFFs policies played an important role in its maintenance over control over the growth, supply, and distribution of rice even at the expense of the community of rice growers. The policies that it has maintained have kept Japanese rice growing in small plots grown by families and individuals who oftentimes have to supplement their income or engage in rice farming part-time because of the economic benefits that farmers reap from rice-growing subsidies. These policies maintained a structure of rice growing that is inefficient and did not reward those rice growers that were most productive. The outcome is that rice growing in Japan has not significantly changed the landscape in nearly 100 years. It is looked upon by outsiders even as a backward industry, that will be unable to compete with large-scale industrial agriculture competition, when that competition is made manifest. It is unfortunate that the timing of programs that encourage full-time farming and consolidation of paddy fields have come so late, occurring only in the past several years, far too late to restructure rice growing in a meaningful way to compete with producers from the United States and other areas even if that was their intended consequence, which is highly questionable. Large-scale industrial agriculture will occur in Japan, it may happen slowly, the demands made by the USA, Europe, and other Asian nations must be answered and Japan has already begun to loosen the reigns on corporate ownership although evidence to illustrate that agricultural corporations are forming with the direction and leadership of foreign enterprises is not there yet. These developments are detrimental for Japan's food security which is already very low. Other policies that are being put forth by the MAFF including the marketing of the traditional Japanese diet to its own population and the world may assist in propping up the demand for rice somewhat, it is unclear yet and it is a question for future researchers to assess the degree of their success.

The pressures from climate change, including the increasing intensity and severity of major storms that affect the global food supply will also play in role in policymaking in Japan on agriculture, it is yet another dynamic to which policymakers must respond. It is predicted that this year a drought in Korea will have an impact on rice growing there which may lead to shortages and an increase in prices. The field of agricultural trade and policymaking is growing increasingly complex because of these changes, and the global political arena with its dynamic shifts in power are part of this picture as well. Future researchers will find a field rich with questions and relationships to examine among these issues and forces. In particular, the issue of genetically modified organisms (GMOs) and the political response to consumer concerns has been understudied to this point. Also, the citizen's environmental movement in Japan responding to crises, consumer issues and food security has been under-researched as well. Although there have been a number of works published on the topic of food security and some articles focusing on Asian food security, there has yet to be a work that examines Japan's food security and its policymaking which addresses both food security and food sovereignty. There has as yet been no work that has updated Dore's work on land reform in Japan, this is a topic that would significantly contribute to researchers understanding of the interplay of weather, climate, soil conditions and the agronomy of Japan, these are all key to understanding the context of rice growing in Japan and provide more detailed information on agriculture itself. While it was not the intention of this book, it is an arena that would be beneficial for future writers on agriculture, agricultural policy, and rice growing in Japan if there were more sources available on these topics written in English. There is a rich literature in Japanese on these subjects, translations of these studies would also be very beneficial for researchers of Japanese agricultural policy who oftentimes confront a dearth of sources in English.

The complex relationships regarding agricultural policy, rice growing, and the social life of Japan are fascinating topics, worthy of research by future scholars. It is my hope that this book provides insight to these academics and perhaps also inspires future academics to build on these humble beginnings that have been the center of my life now for some months.

Appendix: Outline of Introduction, General Provisions, New Basic Law on Food, Agriculture and Rural Areas, Law No. 127

Government of Japan

Introduction: General Provisions

Article 1: Basic Law is intended to develop economy through policies on food, agriculture and rural areas, and to outline the responsibility of state and local governments.

Article 2: Establishes the goal of securing a stable food supply of good quality food at reasonable prices.

Article 3: Discusses the role of agriculture in conservation, water and the environment to form a good landscape as well as importance for maintenance of cultural tradition and as a food supplier.

Article 4: The Sustainable development of agriculture by securing necessary farmlands, irrigation/drainage, a workforce and establishing the structures necessary for the maintenance and improvement of natural cyclical function of agriculture.

Article 5: The improvement in agricultural production conditions and rural welfare, including the living infrastructure, the role of conservation as primary food supplier and multi-functional roles of agriculture.

Article 6: Fisheries and forestry industries will be considered when formulating policy because of the close relationship with food, agriculture and rural areas.

Article 7: The state is responsible for formulating and implementing policies and providing relevant information.

© The Editor(s) (if applicable) and The Author(s) 2019
N. L. Freiner, *Rice and Agricultural Policies in Japan*,
https://doi.org/10.1007/978-3-319-91430-5

Article 8: The local government is responsible for formaulating and implementing policies that suit their natural environment and socio-economic conditions.

Article 9: Farmers and Farmer's organizations are the key constituents in policy.

Article 10: The food industry shall play a role in the efforts to secure a stable food supply.

Article 11: The state and local governments will provide support and coordinate voluntary efforts of farmers, farmers organizations and industry.

Article 12: Consumers are encouraged to have a better understanding of food and be more positive in improving their diets.

Article 13: The state has a legislative responsibility to implement policy through supporting fiscal and financial policy.

Article 13: The Ministry for Agriculture, Fisheries and Forestry will make an annual report to the diet on the state of food, agriculture and rural areas and policies, including which policies were implemented.

BIBLIOGRAPHY

SOURCES

Aldrich, Daniel P. 2008. *Site Fights: Divisive Facilities and Civil Society in Japan and the West*. Ithaca: Cornell University Press.

Agarwal, Vinod K. 2016. Mega-FTAs and the Trade-Security Nexus: The Trans-Pacific Partnership (TPP) and the Regional Comprehensive Economic Partnership. *AsiaPacific Issues* (Analysis from the East West Center) 123: 1–8.

Anderson, Kym. 2017a. *Finishing Global Farm Trade Reform: Implications for Developing Countries*, 6–31. Adelaide, Australia: University of Adelaide Press.

Anderson, Kym. 2017b. Why Open Agricultural Trade Matters. In *Finishing Global Farm Trade Reform: Implications for Developing Countries*, ed. Kym Anderson, 6–31. Adelaide, Australia: University of Adelaide Press.

Anderson, Kym. 2017c. Ongoing and Emerging Issues in Agricultural Trade Negotiations. In *Finishing Global Farm Trade Reform: Implications for Developing Countries*, ed. Kym Anderson, 84–96. Adelaide, Australia: University of Adelaide Press.

Asian Development Bank. 2013. *Food Security in Asia and the Pacific*. Mandayulong, Philippines: Asian Development Bank.

Barrett, Christopher B. (ed.). 2013. *Food Security and Sociopolitical Stability*. Oxford: Oxford University Press.

Bergstand, Jeffrey H. 2016. Should TPP Be Formed? On the Potential Economic, Governance, and Reducing Impacts of the Trans-Pacific Partnership. *East Asian Economic Review* 20 (3): 279–309.

Bestor, Theodore C. 2004. *Tsukiji: The Fish Market at the Center of the World*. Berkeley, CA: University of California Press.

© The Editor(s) (if applicable) and The Author(s) 2019
N. L. Freiner, *Rice and Agricultural Policies in Japan*,
https://doi.org/10.1007/978-3-319-91430-5

Broadbent, Jeffrey. 1998. *Environmental Politics in Japan: Networks of Power and Protest*. Cambridge: Cambridge University Press.

Brown, Lester R. 2004. *Outgrowing the Earth: The Food Security Challenge in an Age of Falling Water Tables and Rising Temperatures*. New York, NY: W. W. Norton.

Bullock, R. 1997. Nokyo. A Short Cultural History. JPRI Working Paper 41. Available at http://www.jpri.org/publications/workingpapers/wp41.html. Accessed October 12, 2017.

Calder, Kent E. 2009. *Pacific Alliance: Reviving US-Japan Relations*. New Haven, CT: Yale University Press.

Cardwell, Michael, and Christopher Rodgers. 2006. Reforming the WTO Legal Order for Agricultural Trade: Issues for European Rural Policy in the Doha Round. *International and Comparative Law Quarterly* 55 (October): 805–838.

Chan, Jennifer. 2008. *Another Japan is Possible: New Social Movements and Global Citizenship Education*. Redwood City, CA: Stanford University Press.

Clever, Jennifer, Midori Iijima, and Benjamin Petlock. 2014. Agricultural Corporations Help Revitalize Japan's Farm Sector. *Global Agricultural Information Network* (GAIN Report #JA4019). Tokyo, Japan: USDA.

Cramer, Gail L., James M. Hansen, and Eric J. Wailes. 1999. Impact of Rice Tariffication on Japan and the World Rice Market. *American Journal of Agricultural Economics* 81 (5). 1149–1156.

Curtis, Gerald. 1989. *The Japanese Way of Politics*. New York: Columbia University Press.

Danahar, Mike. 2003. On the Forest Fringes?: Environmentalism, Left Politics and Feminism in Japan. *Transformations* 6 (February): 1–9.

Davis, Christina L. 2003. *Food Fights Over Free Trade: How International Institutions Promote Agricultural Trade Liberalization*. Princeton, NJ: Princeton University Press.

Davis, Christina, L. and Jennifer Oh. 2007. Repeal of the Rice Laws in Japan: The Role of International Pressure to Overcome Vested Interests. *Comparative Politics* 40 (1): 21–40.

Department of Foreign Affairs and Trade. n.d. Agriculture and Food. In *Japan at the Crossroads: Strategies for the 21st Century*. draft.

Dittmer, Lowell, Haruhiro Fukui, and Peter N. S. Lee (eds.). 2000. *Informal Politics in East Asia*. Cambridge: Cambridge University Press.

Dore, R. P. 1985. *Land Reform in Japan*. New York: Schocken Books.

Eyre, John. D. 1955. Water Controls in a Japanese Irrigation System. *American Geographical Society* 45 (2) April: 197–216.

FAO, IFAD, UNICEF, WFP, and WHO. 2017. *The State of Food Security and Nutrition in the World 2017: Building Resilience for Peace and Food Security*. Rome: FAO.

Fergusoson, Ian F., Mark A. McMinimy, and Brock R. Williams. 2015. The TPP Negotiations and Issues for Congress. *Congressional Research Service.* March 20, CRS.gov. Available at https://fas.org/sgp/crs/row/R42694.pdf. Accessed February 28, 2018.

Francks, Penelope. 2003. Rice for the Masses: Food Policy and the Adoption of Imperial Self-Sufficiency in Early Twentieth-Century Japan. *Japan Forum* 15 (1): 125–146.

Freiner, Nicole. 2013. Mobilizing Mothers: The Fukushima-Daiichi Catastrophe and Environmental Activism in Japan. *ASIA Network Exchange: A Journal for Asian Studies in the Liberal Arts* 21 (1): 27–41. DOI: http://doi.org/10.16995/ane.37.

Freiner, Nicole. 2015a. Japan's Sacred Farmers Brace for Pacific Trade Deals Death Sentence. *The Conversation.com.* Available at https://theconversation.com/japans-sacred-rice-farmers-brace-for-pacific-trade-deals-death-sentence-45280. Accessed October 17, 2017.

Freiner, Nicole. 2015b. Japan May Have Tricky Time Passing TPP After Making Concessions on Rice, Beef. *The Conversation.com* July 30, 2015. Available at https://theconversation.com/japan-may-have-tricky-time-passing-tpp-after-making-concessions-on-rice-beef-48796. Accessed October 17, 2017.

Freiner, Nicole. 2017. Japan's Vote for Abe Could Worsen Prospects for Peace with North Korea, China. *The Conversation.com,* October 23, 2017. Available at https://theconversation.com/japans-vote-for-abe-could-worsen-prospects-for-peace-with-north-korea-china-86173. Accessed February 28, 2018.

Fujibayashi Keiko. 2016. *Japan Instituting Agricultural Reform Programs.* USDA Foreign Agricultural Service, Global Agricultural Information Network, Report JA 6065, December 22, 2016.

Fujibayashi, Keiko. 2017a. *Japan Implements Agricultural Competitiveness Reinforcement Programs.* USDA Foreign Agricultural Service, Global Agricultural Information Network, Report JA 7029, June 26, 2017.

Fujibayashi, Keiko. 2017b. *MAFF 2018 Budget Proposal Shifts Spending to Income Insurance.* USDA Foreign Agricultural Service, Global Agricultural Information Network, Report JA 7114, September 28, 2017.

Fuller, Dorian Q., and Qin Ling. 2009. Water Management and Labor in the Origins and Dispersal of Asian Rice. *World Archaeology* 41 (1) March: 88–111.

Garon, Sheldon. 1997. *Molding Japanese Minds: The State in Everyday Life.* Princeton: Princeton University Press.

General Agreement on Tariffs and Trade (GATT). 1986a. Launching of Uruguay Round. *Focus: GATT Newsletter*, 4I.

GATT. 1986b. *Text of the Agreement.* Available at https://www.wto.org/english/docs_e/legal_e/gatt47.pdf. Accessed December 1, 2017.

GATT. 1993. The Draft Final Act of the Uruguay Round. *The World Economy* 16: 237–260.

George, Aurelia. 1991–1992. The Politics of Interest Representation in the Japanese Diet: The Case of Agriculture. *Pacific Affairs* 64 (4) Winter: 506–528.

George, Timothy S. 2001. *Minamata: Pollution and the Struggle for Democracy in Postwar Japan.* Cambridge: Harvard East Asian Monographs.

George-Mulgan, Aurelia, and Eric Saxon. 1986. The Politics of Agricultural Protection in Japan. In *The Political Economy of Agricultural Protection: East Asia in International Perspective*, ed. Kim Anderson and Yujiro Hayami, 102. Sydney: Allen & Unwin.

George-Mulgan, Aurelia. 2005a. *Japan's Interventionist State: The Role of the MAFF.* London: RoutledgeCurzon.

George-Mulgan, Aurelia. 2005b. Where Tradition Meets Change: Japan's Agricultural Politics in Transition. *The Journal of Japanese Studies* 31 (2, Summer): 261–298.

George-Mulgan, Aurelia. 2006. *Japan's Agricultural Policy Regime.* New York, NY: Routledge.

George-Mulgan, Aurelia. 2008. Japan's FTA Politics and the Problem of Agricultural Trade Liberalisation. *Australian Journal of International Affairs* 62 (2): 164–178.

Gill, Tom. 2013. This Spoiled Soil: Place, People and Community in an Irradiated Village in Fukushima Prefecture. In *Japan Copes with Calamity: Ethnographies of the Earthquake, Tsunami and Nuclear Disasters of March 2011*, ed. Tom Gill, Brigitte Steger, and David H. Slater. Bern, Switzerland: Peter Lang Press.

Gill, Tom, Brigitte Steger, and David H. Slater (eds.). 2013. *Japan Copes with Calamity: Ethnographies of the Earthquake, Tsunami and Nuclear Disasters of March 2011.* Bern, Switzerland: Peter Lang Press.

Godo, Yoshihisa. 2007. *The Puzzle of Small Farming in Japan.* Asia Pacific Economic Papers, No. 365. Canberra, Australia: Australia-Japan Resource Center, ANU College of Asia and the Pacific.

Goodman, Roger (ed.). 2002. *Family and Social Policy in Japan: Anthropological Approaches.* Cambridge, UK: Cambridge University Press.

Gordon, Peter Jegi. 1990. Rice Policy of Japan's LDP: Domestic Trends Toward Agreement. *Asian Survey* 30 (10, October): 943–958.

Hasegawa, Koichi. 2004. *Constructing Civil Society in Japan: Voices of Environmental Movements.* Melbourne, Australia: Trans Pacific Press.

Hayami, Yujiro. 1972. Rice Policy in Japan's Economic Development. *American Journal of Agricultural Economics* (February): 24–31.

Hayami, Yujiro, and Yoshihisa Godo. 1997. Economics and Politics of Rice Policy in Japan: A Perspective on the Uruguay Round. In *Regionalism Versus Multilateral Trade Arrangements, National Bureau of Economic Research*, ed. Takatoshi Ito and Anne O. Kreuger, vol. 6, 371–404. Chicago: University of Chicago Press.

Hori, Chizu. 2015. Ongoing Agricultural Reforms Led by the Abe Administration. *Mizuho Economic Outlook and Analysis.* Tokyo, Japan: Mizuho Research Institute, Ltd.

Horiuchi, Yusaku, and Jun Saito. 2010. Cultivating Rice and Votes: The Institutional Origins of Agricultural Protectionism in Japan. *Journal of East Asian Studies* 10: 425–452.

Ikeda, Yoko. 2013. The Construction of Risk and the Resilience of Fukushima in the Aftermath of the Nuclear Power Plant Accident. In *Japan Copes with Calamity: Ethnographies of the Earthquake, Tsunami and Nuclear Disasters of March 2011*, ed. Tom Gill, Brigitte Steger, and David H. Slater. Bern, Switzerland: Peter Lang Press.

Impoco, Jim. 1992. Japan's Late Greening: Will the World's Self-Ordained Eco-Cop Keep Trashing Its Own Backyard? *US News and World Report* 110 (12) March 16: 61–63.

Interview. 2017a. *By the Author, with MAFF Official.* Accessed July 26.

Interview. 2017b. With Martin Frid, Director, Citizens Union of Japan, Nishoren. Accessed July 28, 2017.

Interviews. 2014. *By the Author with Farmers and JA Members in Hiroshima Prefecture.* Accessed August 1–30.

Interviews. 2017. *By the Author with Farmers and JA Members in Toyama Prefecture.* Accessed July 27–August 1.

Iwamoto, Noriaki. 2003. Local Conceptions of Land and Land Use and the Reform of Japanese Agriculture. In *Farmers and Village Life in Twentieth-Century Japan*, ed. Ann Waswo and Nishida Yoshiaki, 221–243. London: RoutledgeCurzon.

JA Nanto. 2017. JA なんとの概要. Available at www.ja-nanto.jp/company-outline. Accessed December 18, 2017.

JA Tonami/Nanto. 2017. JA となみ野について, 店舗・施設一覧. Available at http://www.ja-tonamino.jp/jainfo/shop. Accessed December 16, 2017.

JA Zenchu. 2017a. JA Multipurpose Cooperative and Activities. Available at www.zenchu-ja.or.jp/eng/multipurpose August 7, 2017. Accessed October 12, 2017.

JA Zenchu. 2017b. JA 全中 について. Available at www.zenchu-ja.or.jp/about/outline. Accessed December 18, 2017.

JA Zen-Noh. 2018. Japan's Cooperatives. Available at www.zennoh.or.jp/english/cooperatives/japancooper.html. Accessed December 18, 2017.

Japan Agricultural News. 2016. Zen-Noh Group and US Producers Band Together for Stable Supplies of Non-GMO Corn. *English Agricultural News*, February 2. Available at http://english.agrinews.co.jp/?p=4216.

Japan Times. 2017. Sowing the Seeds for Lower Food Security? Editorial https://www.japantimes.co.jp/opinion/2017/04/13/editorials/sowing-seeds-lower-food-security/#.WhwYqktJneQ. Accessed November 27, 2017.

Jentzsch, Hanno. 2017. Abandoned Land, Corporate Farming, and Farmland Banks: A Local Perspective on the Process of Deregulating and Redistributing Farmland in Japan. *Contemporary Japan* 29 (1): 31–46.

Johnson, Chalmers. 1986. Tanaka Kakuei, Structural Corruption and the Advent of Machine Politics in Japan. *The Journal of Japanese Studies* 12 (1) Winter: 1–28.

Josling, Tim, David Orden, and Donna Roberts. 2004. *Food Regulation and Trade: Toward a Safe and Open Global System*. Washington, DC: Institute for International Economics.

Jussaume, Raymond A. Jr. 2003. Part-Time Farming and the Structure of Agriculture in Postwar Japan. In *Farmers and Village Life in Twentieth-Century Japan*, ed. Ann Waswo and Nishida Yoshiaki, 199–219. London: RoutledgeCurzon.

Kakuchi, Suvendrini. 2011. Japanese Mothers Rise Up Against Nuclear Power. *The Guardian*, December 22. www.guardian.co.uk.

Kawanaka, N. 1974. Nihon ni okeru seisaku kettai no seiji katei [The Political Process of Policymaking in Japan]. In *Fendai Gyosei to Kanryosei*, ed. Ken Taniuchi, et al. Tokyo: Tokyo University Press.

Kazutoshi, Kase. 2003. Agricultural Public Works and the Changing Mentality of Japanese Farmers in the Postwar Era. In *Farmers and Village Life in Twentieth-Century Japan*, ed. Ann Waswo and Nishida Yoshiaki, 245–266. London: RoutledgeCurzon.

Kikuchi, Masao, and Yujiro Hayami. 1978. Agricultural Growth Against a Land Resource Constraint: A Comparative History of Japan, Taiwan, Korea and the Philippines. *The Journal of Economic History* 38 (4): 839–864.

Krauss, Ellis S., and Robert Pekkanen. 2004. Explaining Party Adaptation to Electoral Reform: The Discreet Charm of the LDP? *Journal of Japanese Studies* 30 (1, Winter): 1–34.

Kumakura, Isao, ed. *Washoku: Traditional Dietary Cultures of the Japanese. Tokyo: Ministry of Agriculture, Fisheries and Forestry (MAFF)*. Available at http://www.maff.go.jp/e/japan_food/washoku/. Accessed December 3, 2017.

Kwa, Aileen, and Walden Bello. 1998. *Guide to the Agreement on Agriculture: Technicalities and Trade Tricks Explained*. Bangkok, Thailand: Focus on the Global South. Available at http://www.jstor.org/stable/10.7864/j.ctt1hfr247.14. Accessed November 30, 2017.

Ladejinsky, Wolf. 1960. Land Reform in Japan. *The Journal of Modern History* 32 (1): 28–31.

Leblanc, Robin M. 1999. *Bicycle Citizens: The Political World of the Japanese Housewife*. Berkeley: University of California Press.

Lewis, Michael. 1990. *Rioters and Citizens: Mass Protest in Imperial Japan*. Berkeley: University of California Press.

Lewis, Michael. 2000. *Becoming Apart: National Power and Local Politics in Toyama, 1868–1945*. Cambridge, MA: Harvard University Press.

Lynch, D., and D. Vogel. 2001. *The Regulation of GMOs in Europe and the United States: A Case-Study of Contemporary European Regulatory Politics.* New York: Council on Foreign Relations.

MAFF. 2007. *New Basic Law on Food, Agriculture and Rural Areas.* Tokyo, Japan: MAFF. Available at http://www.maff.go.jp/j/kanbo/kihyo02/basic_law/pdf/basic_law_agri.pdf. Accessed June 13, 2017.

MAFF. 2015 (April). Summary of the Basic Plan for Food, Agriculture, and Rural Areas: Food, Agriculture and Rural Areas Over the Next 10 Years.

MAFF. 2017. *List of Senior Officials.* MAFF. Available at http://www.maff.go.jp/e/about/official/index.html. Accessed June 13, 2017.

MAFF（農林水産省）. 2017（平成29、年10月）. "米をめぐる状況について." Pamphlet published by MAFF, Tokyo, Japan.

Mason, Robert J. 1999. Whither Japan's Environmental Movement? An Assessment of Problems and Prospects at the National Level. *Pacific Affairs* 72 (2): 187–207.

McCormack, Gavan. 2015. Chauvinist Nationalism in Japan's Schizophrenic State. In *The Politics of the Right: Socialist Register 2016,* ed. Leo Panitch and Greg Albo, 231–249. New York: New York University Press.

McMahon, Paul. 2014. *Feeding Frenzy: Land Grabs, Price Spikes and the World Food Crisis.* Vancouver: Greystone Books.

Miller, Alan S. Reviewer. 1989. Three Reports on Japan and the Global Environment. *Environment* 31 (6): 25–29.

Ministry of Agriculture, Fisheries and Forestry. 1949. *Basic Law on Food, Agriculture, and Rural Areas.* Available at http://www.maff.go.jp/e/policies/law_plan/basiclaw_agri.html. Accessed February 28, 2017.

Ministry of Agriculture, Fisheries and Forestry. 2015. *Vision Statement: Close to Your Daily Life.* Tokyo: Ministry of Agriculture, Fisheries and Forestry.

Ministry of Agriculture, Fisheries and Forestry. 2017. *For Dissemination and Expansion of Good Agricultural Practices (GAP).* Tokyo: Agricultural Production Bureau, MAFF.

Ministry of Agriculture, Fisheries and Forestry (MAFF), Government of Japan. 2006. *Food Safety Basic Law (Tentative Translation).* Law No. 48, May 23, 2003.

Ministry of Agriculture, Fisheries, and Forestry, Government of Japan. 2015 (April). *Summary of the Basic Plan for Food, Agriculture, and Rural Areas: Food, Agriculture and Rural Areas Over the Next 10 Years.* Tokyo: Ministry of Agriculture, Fisheries and Forestry.

Ministry of Agriculture, Fishers and Forestry（農業観光）. 2017. *Overview of Food Compliance Matters.* My Translation. Available at http://www.maff.go.jp/j/syouan/keikaku/beikoku/index.html. Accessed December 15, 2017.

Ministry of Foreign Affairs and Trade, Government of New Zealand. 2016. *Text of Trans-Pacific Partnership Agreement.* Available at http://www.tpp.mfat.govt.nz/tpp-text.php. Accessed November 10, 2017.

Ministry of Health, Labour and Welfare. 2011. *New Standards for Radionuclides in Foods*. Tokyo, Japan: Department of Food Safety, Pharmaceutical & Food Safety Bureau.

Ministry of Justice, Government of Japan. 1952. Agricultural Land Act. Japanese Law Translation Database. Available at http://www.japaneselawtranslation.go.jp/law/detail/?id=2839&vm=04&re=02&new=1. Accessed February 28, 2018.

Mishima, Tokuzoh. 2004. Revision of Japan's Basic Law on Agriculture and Its Features: Improvement of Food Self-Sufficiency Ratio and Agricultural Price Policy. *The Review of Agricultural Economics* 60: 259–271.

Morioka, Rika. 2013. Mother Courage: Women as Activists Between a Passive Populace and a Paralyzed Government. In *Japan Copes with Calamity: Ethnographies of the Earthquake, Tsunami and Nuclear Disasters of March 2011*, ed. Tom Gill, Brigitte Steger, and David H. Slater. Bern, Switzerland: Peter Lang Press.

Munakata, Naoko. 2006. "Has Politics Caught Up with Markets? In Search of East Asian Economic Regionalism". In *Beyond Japan: The Dynamics of East Asian Regionalism*, ed. Peter J. Katzenstein and Takashi Shiraishi, 130–160. Ithaca, NY: Cornell University Press.

Nakai, Yasutaka. 2001. Japan's Clean Development Mechanism and the Fight Against Global Warming. *NIRA Review* (Winter): 12–15.

Nishikawa, Kunio. 2015. *Amendments of the Agricultural Land Act and Measures for Tackling the Idle Farmland Problem (12/27)*. FFTC Agricultural Policy Platform. Food and Fertilizer Technology Center for the Asia Pacific Region. Available at http://ap.fftc.agnet.org/ap_db.php?id=374. Accessed December 4, 2017.

Nye, Joseph. 2004. *Soft Power: The Means to Success in World Politics*. New York: Public Affairs.

OECD. 1990. *Agricultural Policies, Markets and Trade: Monitoring and Outlook*. Paris, France: OECD.

OECD. 2009. *Evaluation of Agricultural Policy Reforms in Japan*. Paris, France: OECD Publishing. Available at https://www.oecd.org/japan/42791674.pdf. Accessed May 16, 2017.

Ohara, Masashi. 2009. *Agriculture in Hokkaido*. Available at https://ocw.hokudai.ac.jp/wp-content/uploads/2016/01/AgricultureInHokkaido-2009-Text-All.pdf. Accessed June 23, 2018.

Ohe, Yasuo. 2011. Evaluating Internalization of Multi-Functionality by Farm Diversification: Evidence from Educational Dairy Farms in Japan. *Journal of Environmental Management* 92: 886–891.

Ohnuki-Tierney, Emiko. 1993. Chapter Four: Rice in Cosmogony and Cosmology. In *Rice as Self: Japanese Identities Through Time*, 44–62. Princeton, NJ: Princeton University Press.

Ohnuki-Tierney, Emiko. 1993b. Chapter Five: Rice as Wealth, Power and Aesthetics. In *Rice as Self: Japanese Identities Through Time*, 63–80. Princeton, NJ: Princeton University Press.

Ohnuki-Tierney, Emiko. 1993c. Chapter Six: Rise as Welf, Rice Paddies as Our Land. In *Rice as Self: Japanese Identities Through Time*, 81–98. Princeton, NJ: Princeton University Press.

Patrick, Hugh. 1976. *Japanese Industrialization and Its Social Consequences*. Berkeley, CA: University of California Press.

Pekkanen, Saadia M. 2005. Bilateralism, Multilateralism, or Regionalism? Japan's Trade Forum Choices. *Journal of East Asian Studies* 5 (1): 77–103.

Pempel, T.J. 2006. A Decade of Political Torpor: When Political Logic Trumps Economic Rationality. In *Beyond Japan: The Dynamics of East Asian Regionalism*, ed. Peter J. Katzenstein and Takashi Shiraishi, 35–62. Ithaca, NY: Cornell University Press.

Prime Minister of Japan. 2015. *Toward an Alliance of Hope*. Address to a Joint Session of Congress, April 29. Available at https://japan.kantei.go.jp/97_abe/statement/201504/uscongress.html. Accessed February 28, 2018.

Pritchard, Bill, and Rebecca Curtis. 2004. The Political Construction of Agro-Food Liberalization in East Asia: Lessons from the Restructuring of Japanese Dairy Provisioning. *Economic Geography* 80 (2) April: 173–190.

Putnam, Robert. 1998. Diplomacy and Domestic Politics: The Logic of Two-Level Games. *International Organization* 42 (Summer): 427–460.

Pye, Lucien. 1985. *Asian Power and Politics. The Cultural Dimensions of Authority*. London, UK: Belknap Press of Harvard University Press.

Rapp, Geoffrey. 2000. Advanced Economic Development, International Trade and Farmers: Is The New Global Economy Bad News for Agricultural Workers? *Drake Journal of Agricultural Law* 5 (Winter): 471–530. Drake University.

Rayner, A. J., K. A. Ingerset, and R. C. Hine. 1993. Agriculture in the Uruguay Round: An Assessment. *The Economic Journal* 103 (421) November: 1513–1527.

Reich, Simon. 2015. Is TPP About Jobs or China?" *The Conversation.com*, May 22, 2015. Available at https://theconversation.com/is-tpp-about-jobs-or-china-42296. Accessed February 28, 2018.

Safecast. 2018a. Documenting Radiation Levels. Available at https://blog.safecast.org/. Accessed May 21, 2017.

Safecast. 2018b. https://blog.safecast.org/. Accessed February 28, 2018.

Sansom, Goerge. 1963. *A History of Japan, 1615–1867*. Stanford, CA: Stanford University Press.

Schanbacher, William D. 2010. *The Politics of Food: The Global Conflict Between Food Security and Food Sovereignty*. Santa Barbara, CA: Praeger.

Schreurs, Miranda. A. 2002. *Environmental Politics in Japan, Germany, and the United States.* Cambridge, UK: Cambridge University Press.

Sherwood, Dave, and Felipe Iturrieta. 2018. Asia-Pacific Nations Sign Sweeping Deal Without the US. *Reuters News,* March 8. Available at www.reuters.com/article/us-trade-tpp/asia-pacific-nations-sign-sweeping-trade-deal-without-u-s-idUSKCN1GK0JM. Accessed June 19, 2018.

Shinoda, Tomohito. 2007. Book Review: Japan's Agricultural Policy Regime by Aurelia George Mulgan. *The Journal of Japanese Studies* 33 (2): 569–573.

Silber, William L. 1985. The Economic Role of Financial Futures. Washington, DC: American Enterprise Institute. Available at http://farmdoc.illinois.edu/irwin/archive/books/Futures-Economic/Futures-Economic_chapter2.pdf. Accessed November 22, 2017.

Slater, David. H. 2011. Fukushima Women Against Nuclear Power: Finding a Voice from Tohoku. *The Asia-Pacific Journal: Japan Focus,* November 9. www.japanfocus.org.

Smil, Vaclav, and Kazuhiko Kobayashi. 2012. *Environmental Impacts: Land, Water, Nitrogen.* Cambridge, MA: MIT Press.

Sohn, Yul. 2010. Japan's New Regionalism: China Shock, Values and the East Asian Community. *Asian Survey* 50 (3) May/June: 497–519.

Solis, Mireya. 2017. Reform/Subsidization Dilemmas in Japanese Trade Policy. In *Dilemmas of a Trading Nation: Japan and the United States in the Evolving Asia-Pacific Order.* Washington, DC: Brookings Institution Press.

Sugiura, Nobuhiko. 2015. Reforming of the Japan Agricultural Cooperatives. *The Japan News by the Yomiuri Shimbun.* Available at http://www.yomiuri.co.jp/adv/chuo/dy/opinion/20150406.html. Accessed September 22, 2017.

Takahashi, Diasuke. 2012. The Distributional Effect of the Rice Policy in Japan, 1986–2010. *Food Policy* 37: 679–689.

Tashiro, Yoichi. 1992. An Environmental Mandate for Rice Self-Sufficiency. *Japan Quarterly* 39 (1): 34–44.

Timmer, Peter C. 2013. Food Security and Sociopolitical Stability in East and Southeast Asia. In *Food Security and Sociopolitical Stability,* ed. Christopher B. Barrett, 452–474. Oxford: Oxford University Press.

Tsutsui, Masao. 2003. The Impact of the Local Improvement Movement on Farmers and Rural Communities. In *Farmers and Village Life in Twentieth-Century Japan,* ed. Ann Waswo and Nishida Yoshiaki, 60–78. London: RoutledgeCurzon.

UNESCO. 2013. *Washoku, Traditional Dietary Cultures of the Japanese, Notably for the Celebration of New Year.* Available at https://ich.unesco.org/en/RL/washoku-traditional-dietary-cultures-of-the-japanese-notably-for-the-celebration-of-new-year-00869. Accessed June 21, 2018.

UNFAO. 2018. *Japan Country Profile.* Available at http://www.fao.org/countryprofiles/index/en/?iso3=JPN. Accessed June 21, 2018.

UNFAO, FAOSTAT Food, and Agriculture Data. http://www.fao.org/faostat/en/#home. Accessed November 18, 2017.

Verba, Sidney, Norman H. Nye, and Jae-On Kim. 1971. *The Modes of Democratic Participation: A Cross National Comparison.* Beverly Hills, CA: Sage Publications.

Waswo, Ann. 2003. In Search of Equity: Japanese Tenant Unions in the 1920s. In *Farmers and Village Life in Twentieth-Century Japan*, ed. Ann Waswo and Nishida Yoshiaki, 79–125. London: RoutledgeCurzon.

Waswo, Ann, and Nishida Yoshiaki (eds.). 2003. *Farmers and Village Life in Twentieth-Century Japan.* London: RoutledgeCurzon.

Yoshiaki, Nishida. 2003. Dimensions of Change in Twentieth-Century Rural Japan. In *Farmers and Village Life in Twentieth-Century Japan*, ed. Ann Waswo and Nishida Yoshiaki, 38–59. London: RoutledgeCurzon.

Yoshida, Reiji and Mizuho Aoki. 2015. Abe's JA Reforms No Salve for Core Problems, Experts Say. *The Japan Times* (February 19).

Yoshimatsu, Hidetaka. 1998. Japan's Keidanren and Political Influence on Market Liberalization. *Asian Survey* 31 (3, March): 328–345.

Yoshimatsu, Hidetaka. 2007. Japan's Quest for Free Trade Agreements Constraints from Bureaucratic and Interest Group Politics. In *Japan's Future in East Asia and the Pacific*, ed. Mari Pangestu and Ligang Song, 80–102. Canberra: The Australian National University Press.

INDEX

© The Editor(s) (if applicable) and The Author(s) 2019
N. L. Freiner, *Rice and Agricultural Policies in Japan*,
https://doi.org/10.1007/978-3-319-91430-5